Thermoregulation of Sick and Low Birth Weight Neonates

Springer
Berlin
Heidelberg
New York
Barcelona
Budapest
Hong Kong
London
Milan
Paris
Santa Clara
Singapore
Tokyo

Albert Okken · Jochim Koch (Eds.)

Thermoregulation of Sick and Low Birth Weight Neonates

Temperature Control
Temperature Monitoring
Thermal Environment

With 87 Figures

Springer

Prof. Dr. Albert Okken
Wilhelmina Childrens Hospital
P.O. Box 18009
3501 CA Utrecht
The Netherlands

Dr. Jochim Koch
Drägerwerk AG
Moislinger Allee 53/55
23542 Lübeck
Germany

Cataloging-in-Publication Data applied for
Die Deutsche Bibliothek - CIP-Einheitsaufnahme
Thermoregulation of sick and low birth weight neonates: temperature control, temperature monitoring, thermal environment / Albert Okken; Jochim Koch. - Berlin; Heidelberg; New York; London; Paris; Tokyo; Springer, 1995

ISBN-13: 978-3-642-79936-5 e-ISBN-13: 978-3-642-79934-1
DOI: 10.1007/978-3-642-79934-1

NE: Okken, Albert; Koch Jochim

Softcover reprint of the hardcover 1st edition 1995

Typesetting: Camera-ready by authors

SPIN: 10488501 26/3134 - 5 4 3 2 1 0 - Printed on acid-free paper

Preface

Over the past 15 years, the population of sick neonates has changed. The average birthweight has decreased dramatically to as low as 450g while gestational age has been reduced to 22-24 weeks. These babies have totally different needs and require more complex health care. It is now time to change previous recommendations which were based on more mature neonates.

In this foreword we aim to give some of the background into the ideas which lead to us starting work on new guidelines. The question was : Can the thermal condition of the baby be better monitored and investigated and, if so, how ?

Measurement of body temperature remains a very important method of ascertaining the thermal condition of a newborn baby. There is an implicit international agreement that the body temperature of an infant is measured either intermittently using a liquid expansion thermometer or continuously by means of temperature sensors placed in the rectum or the axilla.

Rectal temperature is not necessarily identical to the core temperature. There may be up to a 2 °C difference and this could result in an incorrect assessment of the condition of the patient. For example in the presence of cold lower limbs, the cool venous blood returning from the legs will flow close to the rectum and, as a result, the rectal temperature may be significantly lower than that of the core. Other workers have shown how rectal temperature depends on the depth to which the sensor has been inserted.

Furthermore, core temperature should not be considered as being uniform throughout the body. There are differences in the temperature of individual organs such as brain, liver, kidney, heart and lungs. This may be a reflection of heat production in some organs (brain, liver) or heat loss in others (lungs), or due to loss of heat from the blood on its path to the organs. This problem may be avoided by using a 'local core temperature' such as that of the oesophagus.

The adjustment of the incubator climate (temperature and humidity) for the individual baby still causes problems for the nursing staff. Attempts to use core temperature as a means of determining settings can result in an incorrect assessment of the actual thermal condition of the baby.

Simultaneous measurement of the difference between a central (rectal) and peripheral (sole of foot) temperature may give more useful information :

Nursing procedures result in significant increases in the central-peripheral temperature gap. The severity of this change depends on body weight with the major effects seen in infants below 1100g. Continuous display of the changing temperatures to the nursing staff results in them altering practice to reduce increases in the central-peripheral gap. The thermal and metabolic effects of a changing temperature gap are discussed briefly.

Changes in the central-peripheral temperature gap are also useful in assessing sepsis in the newborn, with a diagnosis of septicaemia possible before culture results are available.

Knowledge of simultaneous changes in a central and a peripheral temperature therefore provides the paediatrician with better information than can be obtained from a single measurement. By monitoring these temperatures continuously and displaying them as time trends, along with other parameters such as heart rate and blood pressure, it is possible to detect a number of conditions in the baby : Centralisation of blood due to peripheral vaso-constriction can be detected. This is an early sign of cold stress and hypovolaemic shock, with the latter being diagnosed by the associated changes in heart rate and, at a later stage, blood pressure. These conditions can be distinguished from pyrexia, hyperthermia and sepsis.

Other workers throughout the world are in agreement that the measurement of temperature at two sites results in a better assessment of the condition of the baby and gives better information on which nursing staff can judge the adjustment of the thermal environment. A full evaluation of the baby relies on the measurement of other physiological parameters, such as heart rate, blood pressure and oxygen saturation, as well as the results of physical examination.

Provision of appropriate temperature monitors and the training of nursing staff to minimise changes in the central-peripheral temperature difference should improve the stability of preterm babies and reduce morbidity. The introduction of a means of continuously measuring core temperature will improve the reliability of thermal monitoring. Greater stability of the patient may enable nursing time to be redirected to other aspects of care. The studies reported here show how the measurement of two body temperatures, and their display as trend graphs, lead to significant improvements in the quality of thermal care of the newborn, and in particular the preterm infant.

The work resulting in these guidelines was presented and discussed by all the authors at a workshop in Schloß Reinhartshausen, Germany, October 1994. The discussions proved very fruitful and the results are now published in this book. They will allow us to improve the quality of health care for all sick and low birthweight infants.

Albert Okken
Jochim Koch

Contributors

Dimitris Anagnostakis,
Associate Professor of Paediatrics
First Department of Paediatrics,
Athens University School of Medicine,
"Aghia Sophia" Children's Hospital, Athens / Greece

Karl Bauer,
Paediatrican
Freie Universität Berlin, Universitätsklinikum Steglitz,
Berlin / Germany

Stephen Baumgart,
Professor of Paediatrics
Thomas Jefferson University, Neonatology Division,
Philadelphia PA / USA

Edward F. Bell,
Professor of Paediatrics
University of Iowa, Department of Paediatrics,
Iowa City, Iowa / USA

Karl Henrik Fridblom,
Anestesiologist
Visby Hospital
Visby / Gotland / Sweden

Jane Hailey,
Paediatric Registrar
Dept. of Child Life and Health, The University of Edinburgh,
Edinburgh / Scotland / UK

Jochim Koch,
Mechanical Engineer
Drägerwerk AG,
Lübeck / Germany

Peter Lemburg,
Professor of Paediatrics
Medizinische Einrichtungen der Heinrich Heine-Universität,
Zentrum für Kinderheilkunde
Neonatologie und Pädiatrische Intensivmedizin,
Düsseldorf / Germany

Andrew Lyon,
Consultant Neonatologist
Simpson Memorial Maternity Pavilion
Edinburgh / Scotland / UK

Heike Martiny,
Professor of Microbiology
Freie Universität Berlin, Institut für Hygiene,
Berlin / Germany

Neil McIntosh,
Professor of Paediatrics
Dept. of Child Life and Health, The University of Edinburgh,
Edinburgh / Scotland / UK

Johannes Messaritakis,
Professor of Paediatrics
First Department of Paediatrics,
Athens University School of Medicine,
"Aghia Sophia" Children's Hospital, Athens / Greece

Jean L. Micheli,
Paediatrican
Dept. of Pediatrics, CHUV-University Hospital,
Lausanne / Switzerland

Albert Okken,
Professor of Paediatrics
Wilhelmina Childrens Hospital
Utrecht / The Netherlands

Madeline Pikaar,
Research Fellow
Dept. of Child Life and Health, The University of Edinburgh,
Edinburgh / Scotland / UK

Pieter J. J. Sauer,
Professor of Neonatology
Afd. Neonatologie, Sophia Kinderziekenhuis,
Rotterdam / The Netherlands

Gunnar Sedin,
Professor of Perinatal Medicine
Dept. of Paediatrics, University Children's Hospital,
Uppsala / Sweden

Georg Simbruner,
Professor of Paediatrics
Dr. v. Haunersches Kinderspital, Ludwig Maximilians University,
München / Germany

John C. Sinclair,
Professor of Paediatrics and of Clinical Epidemiology and Biostatistics
Dept. of Pediatrics, McMaster University, Medical Center,
Hamilton, Ontario / Canada

Ragnar Tunell,
Emeritus Paediatrican
Karolinska Institut, Dept of Pediatrics, Huddinge Hospital,
Huddinge / Sweden

Hans Versmold,
Professor of Paediatrics
Freie Universität Berlin, Universitätsklinikum Steglitz,
Berlin / Germany

Andrew Wilmshurst
Paediatrics Registrar
Dept. of Child Life and Health, The University of Edinburgh,
Edinburgh / Scotland / UK

Contents

Thermodynamic principles

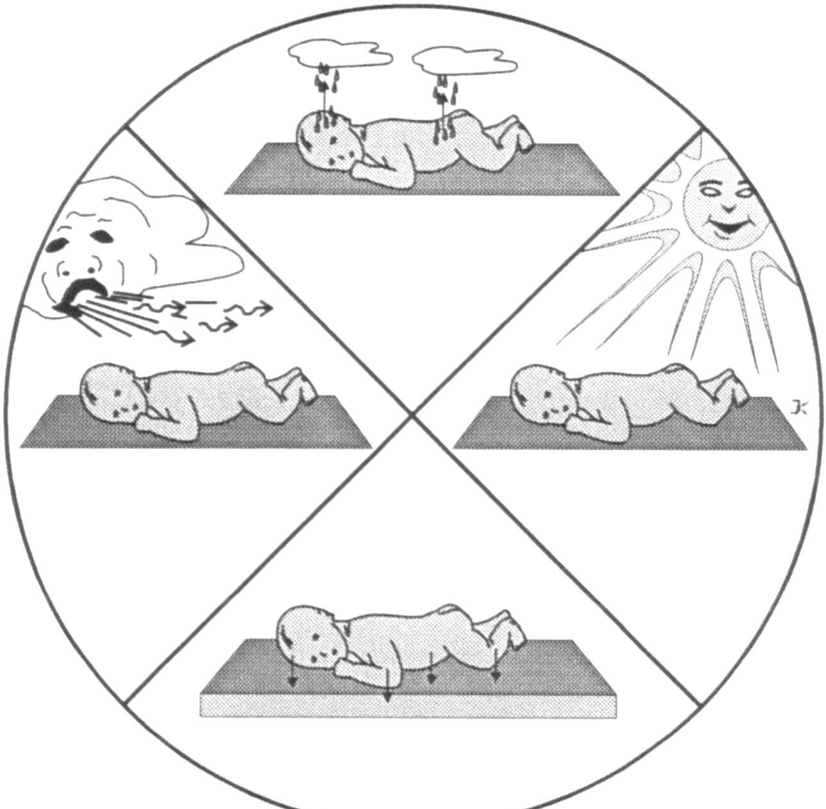

Chapter 1

The Concept of Thermoregulation

ALBERT OKKEN

Introduction

Bodytemperature is precisely controlled within certain limits at a level optimal for normal function and survival. From many studies in animal and man the model of thermoregulation has been developed. In man, most data about the function of the thermoregulatory system have been obtained from studies in adults and to some extent in the fullterm neonate. Much less is known about the function of the system in preterm very low birthweight infants. This chapter will discuss elements of thermoregulation in relation to prematurity of human newborn infants.

The Thermoregulatory Model

Control of bodytemperature is achieved by a complex system which via negative feed-back basically creates a balance between heat production and heat gain and heat loss. A schematic representation of the thermoregulatory system is given in figure 1.-1. The key of this system is a central controller located in the hypothalamus and limbic system, which based on information from central- and peripheral thermoreceptors (multiple-input) controls the action of the so-called effectors: thermogenesis, the vasomotor system, sweat secretion and thermoregulatory behaviour, through the efferent nervous system. Bodytemperature therefore is the result of the combined action of the detectors, controller system and the effectors. An excellent review of the thermoregulatory system in neonatal animal and man is given in reference 1.

Disturbances in one or more of the elements of thermoregulation will result in an abnormal bodytemperature. In case of newborn infants, especially preterm infants, immaturity of the thermoregulatory system makes the infant more vulnerable for changes of environmental temperature.

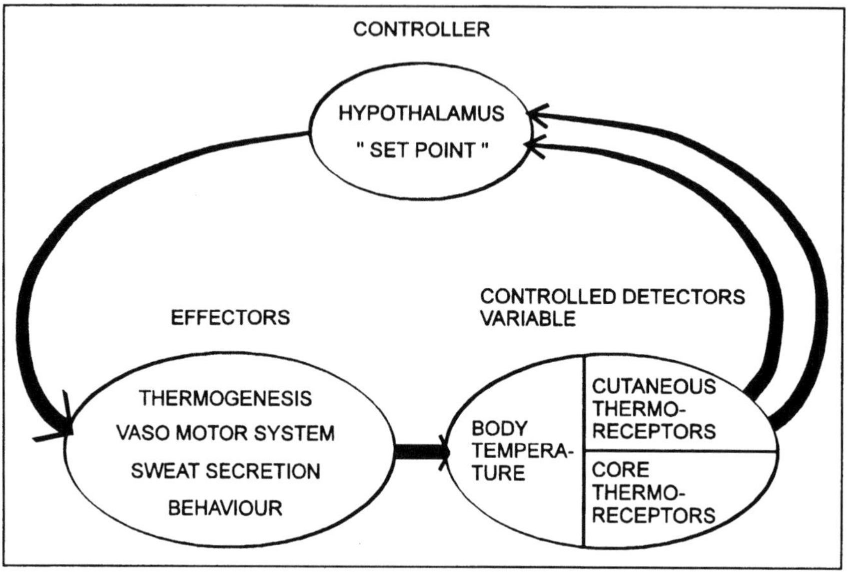

Fig. 1.-1 : Schematic presentation of the thermoregulatory system

Controlled Variable and Controller; Bodytemperature and "Hypothalamic" Temperature

In the body, heat is produced by various tissues and organs with different metabolic rates. Therefore various temperatures at different sites in the body can be found. In newborn infants, the contribution of different organs relative to total body mass is quite different from adults.

The contribution of the brain to heat production is approximately 6 times higher (Table 1.-1), of the liver 2 times and in contrast to the adult, of the skeletal muscles only half. This means that in the newborn a significant part of heat is produced in the most upper part of the body next to the controller. Interestingly body "core" temperature in most studies and in clinical care is measured in the lower part of the body (rectal temperature) or in the axillar or at the abdominal skin. There is no doubt that monitoring of bodytemperature at those sites has proven its value in clinical care, but its significance in studies of regulation of temperature has to be questioned. Single point measurements such as rectal temperature and axillary temperature may underestimate body core temperature in a cool environment [2]. A few studies have reported "normal" bodytemperature in preterm- and fullterm newborn infants [3, 4, 5]. Although slight differences are reported, "normal" bodytemperature ranges around 37 °C.

Tab. 1.-1 : Role of effectors; heat production of the brain

BODYWEIGHT (KG)	BRAINWEIGHT (G) (% B.W.)	HEATPRODUCTION (%)
1.1	190 (17)	?
3.5	475 (14)	60 - 80
5.5	650 (12)	50 - 70
70.0	1400 (2)	23

Effectors

Thermogenesis

Thermoregulatory Thermogenesis

In full term newborn infants, non-shivering thermogenesis (oxidation of brown adipose tissue) is the major route of a rapid increase of heat production in response to cold exposure. Increased activity by increasing body-movements has also been reported. During the first year of life the role of shivering thermogenesis increases and of non-shivering thermogenesis decreases. In very preterm infants, brown adipose tissue can be identified around 26 weeks gestational age. The amount of brown adipose tissue increases rapidly to about 10 % of total body adipose tissue near term [6]. The heat produced in brown adipose tissue is the result of the activity of a unique mitochondrial uncoupling protein "thermogenin" [7]. Brown tissue oxidation is activated by the hypothalamus (controller system) via the sympathic nervous system. Very preterm infants have no ability to oxidize substantial amounts of brown adipose tissue which limits their thermo-regulatory thermogenesis.

Vasomotor System

Circulatory Thermoregulation

Thermoregulatory vasoconstriction and vasodilatation are important components of the transportation of heat throughout the body and subsequent heat loss. Studies have shown that peripheral blood flow is affected by environmental temperature in both term and preterm newborn infants [8, 9]. In the presence of a subcutaneous layer of fat peripheral vasoconstriction can result in a substantial reduction of heat loss. This is the case in fullterm infants. In very preterm infants (below 26 weeks gestation) the subcutaneous fat layer is very thin which limits the reduction of heat loss through vasoconstriction considerably.

Sweat Secretion

In a warm environment the production of sweat is a way to increase heat loss. The innervation of sweat production is (like vasodilatation) regulated by the hypothalamus (controller) through the sympathic nervous system. Fullterm newborn infants have more sweat glands compared to adults but their response is only one third of those of adults [10, 11]. In a warm environment not all newborns start to sweat possibly via different responses of the controller system [12].

The role of effectors which mostly control bodytemperature in a cool or in a warm environment is summarized in table 1.-2.

Tab. 1.-2 : Role of effectors in the adult, full term and preterm infants

EFFECTOR	ADULT	FULL-TERM	PRETERM
SHIVERING THERMOGENESIS	+	± or -	-
NON-SHIVERING THERMOGENESIS	+	+	± or -
VASOMOTOR SYSTEM (CUTANEOUS BLOOD FLOW)	+	+	+ to ±
SWEAT SECRETION	+	+ or ±	-

Implication for Practice

It is general practise to keep the bodytemperature of newborn infants within the range normal for adults. For very preterm infants intra-uterine bodytemperatures are not recommended. Most studies in newborn infants focus on the function of the effectors. In very preterm infants, the effectors have a very limited function.

Recommendations for Future Research

The following items should be re-evaluated for very low birthweight newborn infants:

- Reference values for normal bodytemperature
- The site(s) representing core temperature
- Site(s) and reference values for clinical monitoring
- Postnatal development of thermoregulation of very low birthweight newborn infants needs further attention

References:

1. Brück K (1992) Neonatal thermal regulation. In: Polin R, Fox WW Fetal and neonatal physiology, Philadelphia USA, pp 488-515.

2. Chelappah G (1980) Aspects of thermoregulation in term and preterm newborn babies. Thesis University of Nottingham England.

3. Wheldon AE, Hull D (1983) Incubation of very immature infants. Arch Dis Child. 58:504-508.

4. Bell EF, Rios GR (1983) Air versus skin temperature servo control of infants in incubators. J Peds 103:954-959.

5. Mayfield SR, et all (1984) Temperature measurements in term and preterm neonates. J Peds 104:271-275.

6. Trayhurn P, Micholls DG, eds. (1986) Brown adipos tissue. London, Edward Arnold Ltd.

7. Klingenberg M (1990) Mechanism and evolution of the uncoupling protein of brown adipose tissue. TIBS 15:100-112.

8. Hey EN, Katz G (1970) The range of thermal insulation in the tissues of the newborn baby. J Physiol 207:667-681.

9. Beaufort-Krol GCM et al (1989) Postocclusive reactive hyperaemia of cutaneous blood flow in premature newborn babies. Act Paedr Scand Suppl 360:20-24.

10. Foster KG, Hey EN, Katz G (1969) The response of the sweat glands of the newborn baby to thermal stimuli and to intradermal acetylcholine. J Appl Physiol 203:13-29.

11. Sulyok E, Jéquier E, Prod'hom LS (1973) Thermal balance of the newborn infant in a heatgaining environment. Pediatr Res 7:888-900.

12. Bach V, Bouferrace B, Kremp O et al (1994) Regulation of sleep and bodytemperature in response to exposure to cool and warm environments in neonates. Pediatr Res 93:789-796.

Chapter 2

Metabolic Background of Neonatal Heat Production, Energy Balance, Metabolic Response to Heat and Cold

PIETER J. J. SAUER

Introduction

It has been known for a long time that there is an association between survival of the newborn and environmental temperature, particularly when the infant is of low birthweight. The 17th century story of the low birthweight infant nursed next to an oven in a bakery is well known. Budin, writing at the end of the last century, made important contributions to our understanding of the relationship between environmental temperature and survival. He showed that mortality was lower if infants had higher body temperatures and he started to use incubators to keep babies warm [1]. Most of the studies on thermoregulation of the newborn have been conducted since the second world war with the early ones confirming the relationship between environmental temperature and survival [2]. Later studies found also an association with growth [3], with those infants nursed at a higher environmental temperature, the so called normal core temperature, showing a higher weight gain when compared to those at lower body temperatures. It is noteworthy that the "normal body temperature" of 36 °C or less used in these studies was lower than that considered normal today [4].

Animal studies

Studies in different species have addressed the response of the animal to cooling, both before and after birth, but care must be taken when drawing conclusions. Three different types of animals must be distinguished :

1. In precocial newborns (eg sheep, guinea pigs) the demand for heat production is often maximal at birth and cold stress or a noradrenaline infusion results in an immediate large metabolic response.

2. Altricial newborns (eg rats) are born naked in a large litter and huddle together in a nest. A response to noradrenaline develops progressively in the days after birth.

3. Immature newborns appear to lack a thermoregulatory capacity at birth and this develops from about 10 days of age.

To which category does the human infant belong ? Although the term infant is able to increase oxygen consumption in response to a cool environment, he can also change posture and can show a reduction in body temperature. Moreover the capacity to produce heat is not maximal at birth. There are few studies in preterm infants and these show that although there is a rise in oxygen consumption in response to a cool environment, the drop in body temperature is more marked than in term babies [5, 6, 7]. Data on the effects of cooling in the extremely low birthweight infant of short gestation are lacking.

Brown Adipose Tissue (BAT)

In response to a cool environment the newborn animal has two options to increase heat production and prevent a drop in body temperature - shivering or activity and non-shivering thermogenesis.

Human infants become active and start crying when their temperature falls by 1-2 °C but a first response is non-shivering thermogenesis which takes place in many organs of the body. The most important site is brown adipose tissue which increases during fetal life and has a structure that is specialised for heat production. It contains a high concentration of stored triglycerides, a rich capillary network and is densely innervated with sympathetic nerve endings on the vessels and on each adipocyte. Each cell has numerous mitochondria with respiratory chain enzymes but the unique property is the uncoupling protein which is the rate limiting enzyme in the process of heat production [8]. The action of this protein means that when fat is oxidised heat is produced rather than energy rich phosphate bonds which is the usual result in other tissues. Noradrenaline stimulates lipolysis and the activity of the uncoupling protein and thus produces heat. Thyroxine plays an important role, being taken up by the cell and converted to the active T3 by the enzyme Type II 5' deiodinase which stimulates the uncoupling protein. This role of thyroxine has been studies in many animals. Thyroidectomy at mid gestation in the fetal lamb results in profound hypothermia at birth with the animals dying soon after [10]. Removal of the thyroid towards the end of gestation caused a reduction in rectal temperature and lower FFA levels in the newborn [11], while thyroidectomy at delivery abolished the postnatal surge in plasma T3 without any effect on body temperature or oxygen consumption [11, 12]. An infusion of T3 caused no increase in non-shivering thermogenesis in the lamb in-utero [13] while, in contrast, T4 infusion resulted in a normalisation of uncoupling protein in the hypothyroid rat [14]. A recent study using rats also failed to show a role for T3 in raising metabolic rate as a response to cooling [15] and it seems that the conversion of T4 to T3 is crucial for the thermogenic response. The acute surge in thyroid hormones (T3 and T4) at birth appears to be of limited significance with regard to neonatal thermogenesis. The effect of noradrenaline and the intracellular conversion of T4 to T3 that occurs at birth seems to be of more importance.

Initiation of Thermogenesis

The temperature of the fetus follows, and is approximately 0.5 °C higher than, that of the mother. Directly after the birth the newborn has to regulate his own temperature. Gunn and Gluck have conducted interesting studies into the initiation of thermogenesis in pregnant sheep. Cooling the fetus in-utero caused a very limited increase in FFA and glycerol, while the body temperature dropped significantly [16, 17]. Ventilation of the fetus, thereby increasing the oxygen tension, caused a slightly greater increase in FFA and glycerol and a modest rise in body temperature. Clamping the umbilical cord had a profound effect with a sharp increase in FFA and glycerol and a rapid rise in temperature. Removing the clamp caused a reduction in FFA and a drop in body temperature. These studies suggest that there is an inhibitor of thermogenesis before birth, probably produced by the placenta. Its exact nature has not been established but potential candidates are adenosine and prostaglandin E.

Hypoxia

Hypoxia impairs non-shivering thermogenesis [16] but studies are unethical to perform on human infants and are difficult in newborn animals who may not survive the investigation. Animals subjected to hypoxia after birth are often anaesthetised making the conclusions on thermogenesis difficult to interpret. The shivering and non-shivering thermogenesis in 2 month old rats before and after carotid denervation has been recently studied [18] and it was concluded that :

1. the interaction between environmental temperature and oxygen content in the air on the control of respiration is partly dependent on the carotid body afferents,

2. shivering thermogenesis may be transiently affected by hypoxia independently of the carotid body afferents and

3. non-shivering thermogenesis may be directly inhibited by hypoxia, especially during exposure to cold.

This last conclusion is especially important as it implies that the newborn subjected to the combination of hypoxia and cold stress is at high risk of decreasing body temperature. More studies in the newborn are needed but this conclusion agrees with clinical experience.

Substrate for Brown Adipose Tissue

The substrate for fat oxidation within brown adipose tissue (BAT) is unknown. As BAT contains lipids, fatty acids may be used but, as BAT is not readily depleted, it is likely that it takes up fuel from the plasma. This can be either fatty acids or glucose. A study in newborn pigs showed that fat administered with colostrum was used for thermogenesis at a low environmental temperature [19]. However the liver glycogen stores were more depleted in pigs receiving low fat compared to normal colostrum, suggesting that glucose was also used as a substrate.

Studies have looked at the glucose transporters in different organs, including BAT. GLUT1 is mainly responsible for the regulation of glucose uptake under basal conditions, while GLUT4 controls the increase in glucose uptake in response to insulin. The expression of GLUT1 and GLUT4 genes in skeletal muscle, heart and BAT is developmentally regulated, with GLUT1 being the predominant glucose transporter in muscle and BAT during fetal and early neonatal life [20].

The high activity of GLUT1 in BAT before birth may indicate a high glucose uptake and this may be an important energy store for this tissue. After birth there is a fall in GLUT1 expression followed by a rise a few days later. This may indicate that energy stores are depleted after a few days and then replenished from glucose, but at present this is pure speculation. There is also a rise in GLUT4 expression in BAT soon after birth and this, somewhat surprisingly, appears to be augmented by hypothyroidism [21]. The exact role of glucose in BAT is therefore still unknown. When glucose is taken up into BAT, converted to fat and subsequently oxidised, it will be measured by indirect calorimetry as glucose oxidation. Therefore a high RQ during the first hours of life does not rule out an important role for glucose as an energy source for BAT.

Exposure to Low Environmental Temperatures, Human Studies

The question "How does a human newborn infant react to environmental temperatures below the so called neutral range ?" is not easy to answer using existing data and studies are difficult to perform. It is not ethical to expose a newborn to a cool environment and data from accidental exposure are difficult to interpret. Some statements can be made by combining the results of a number of studies. Exposure of infants to an environmental temperature several degrees below the so called neutral temperature causes an increase in oxygen consumption and therefore heat production [22, 23], although this is seldom enough to prevent a fall in body temperature. Heat loss in naked infants of around 2 kg was, in one older study, linearly related to environmental temperature with a value of approximately 35 kcal/m^2.hr at

30 °C on the 2nd postnatal day [24]. This loss is much higher than the heat production of 17 kcal/m^2.hr in a thermoneutral environment and the infant will have to double its heat production to compensate. This is impossible resulting in a drop in body temperature. Studies done in low birthweight babies show them to react with a steeper drop in body temperature [26] but there was no clear distinction made between AGA and SGA infants. Oxygen consumption of low birthweight infants is less at all environmental temperatures. The rapid drop in body temperature is further explained by their higher surface/weight ratio, poorer body insulation and lower stores of fatty acids (figure 2.-1) when compared with infants born at term.

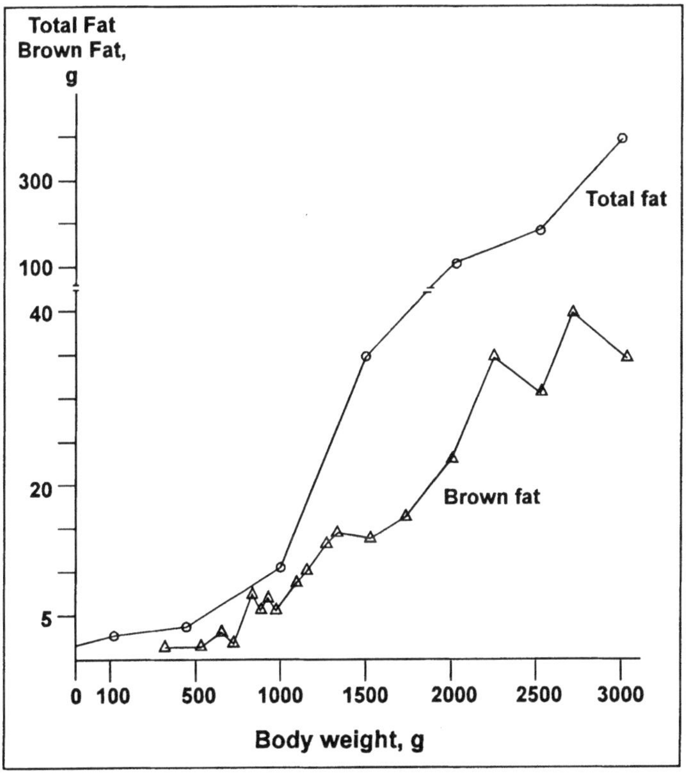

Fig. 2.-1 : The weights of total fat and brown fat are plotted against body weight of the human fetus [32].

The small change in environmental temperature may have a different effect on the body temperature and oxygen consumption of the low birthweight infant when compared with that of a larger change. We and others observed an alteration in body temperature without any change in oxygen consumption in infants exposed to an environmental change of 1 °C [23, 26]. It may be that the infant is reacting to a small change in environmental temperature with a physiological response - reduced skin blood flow, small drop in core temperature and change in posture. The infant will increase heat production as a response to a larger change in environmental temperature.

It is likely that both responses are dependent on gestational and postnatal age as well as the availability of substrates for heat production. Although it is well known that extremely low birthweight infants are very prone to hypothermia, no data are available on their ability to either reduce heat loss or increase heat production.

Heat Production for Thermogenesis

Increases in heat production are obtained from shivering or non-shivering mechanisms. Newborn infants do not appear to really shiver but become restless and active when exposed to cold. Non-shivering thermogenesis is an important form of thermogenesis in the newborn.

Lean demonstrated in 1986 the presence of uncoupling protein in the brown adipose tissue of all infants [27] although the levels are lower in the preterm compared with term. A recent study has shown that the structure of BAT is well developed in the preterm infant even at gestational ages as low as 25 weeks. The role of BAT was studied in 13 newborns, of gestational age 25-40 weeks, who survived at least 4 days [28]. The specific content of uncoupling protein increased from 29.4 ± 3.3 pmol/mg protein at 25 weeks to 62.5 ± 10.2 pmol/mg at 40 weeks. The uncoupling protein/F1-ATPase molar ratio, a sensitive marker of brown fat differentiation, also increased with gestation. This study also showed the importance of the intracellular conversion of T4 to T3 by the enzyme type II 5'-deiodinase in the activity of BAT. The enzyme was active by 25 weeks gestation and this increased fourfold towards term. The major increase in uncoupling protein and deiodinase activity was observed before 32 weeks. These results support the hypothesis that non-shivering thermogenesis is important in newborn infants, both term and preterm, but the level of cold stress at which these processes become active is unknown.

Oxygen Consumption soon after Birth

Methodological constraints make the measurement of oxygen consumption difficult in the first minutes and hours after birth. The use of supplemental oxygen causes problems for most calorimeters [29]. Energy expenditure in stable preterm infants, nursed at a neutral temperature, increases gradually in the first week of life from approximately 40 kcal/Kg/d on day 1 to 60 kcal/kg/d after a week [23]. Metabolic rate is influenced by a number of factors including energy intake. A study in the first few minutes to hours after birth showed higher values for oxygen consumption of up to 10 ml/Kg.min, equivalent to an energy expenditure of approximately 80 kcal/kg/d, than found in infants at a few days of life [30]. These higher values are also seen in infants with respiratory distress syndrome and therefore may be related to the high work of breathing immediately after birth [31].

Substrate Utilisation in the First Days of Life

Substrate utilisation can be estimated either from the levels of different compounds in plasma, or from the respiratory quotient as measured from indirect calorimetry or from ^{13}C labelled compounds. Heim used plasma values during the first days of life and during the response to hypothermia [32], and he showed a decrease in body temperature and increase in oxygen consumption as a response to a cool environment. The drop in body temperature was more marked in the preterm infant while the increase in oxygen consumption was less. However the observed RQs were very low, around 0.7, indicating that only fat was being oxidised both during exposure to a normal environment and hypothermia. This last finding casts doubt on the validity of the results. Blood glucose, lactate, plasma free fatty acid and glycerol levels were also measured as indicators of substrate consumption. Plasma free fatty acids and glycerol increased during cold exposure, both in infants weighing 3-4 kg as well as those 1-2 kg. Glucose showed no marked changes while lactate increased, particularly in the tiny babies. Infants were able to increase oxygen consumption even during hypoglycaemia, but the infusion of glucose caused a rapid fall in oxygen consumption [32].

From all these results one can speculate that glucose and fatty acids are both used by the newborn infant during the first days of life, whether in a normal thermal environment or during hypothermia (Figure 2.-2)

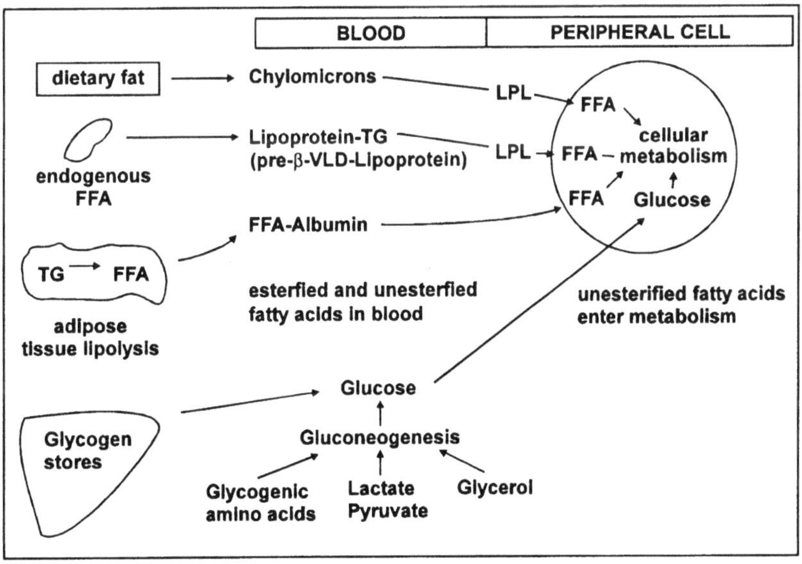

Fig. 2.-2 : The main energy sources for cold-inducted thermogenesis in newborn infant [32].

The RQ of the preterm infant in the days after birth is above 0.9 indicating that glucose is the major energy source but one must be careful about the use of this method of estimating substrate utilisation. In studies using both RQ and ^{13}C labelled glucose, differences in glucose oxidation measured by each method are observed at RQs below 0.8 and above 0.9. Possible explanations for these differences are discussed elsewhere [33]. One must be careful therefore before concluding, from the results of indirect calorimetry, that glucose is the main source of energy during the first days of life.

We have studied glucose oxidation, calculated from the $^{13}CO_2$ in expired air following an infusion of U-^{13}C-glucose, soon after birth in ventilated preterm infants [34]. An average value of 4.1 mg/kg.min, equal to approximately 20 kcal/kg.d, was found. Assuming a total energy expenditure of around 45 kcal/kg.d, this indicates that the preterm infant is using other sources of energy during the first days of life. Using ^{13}C-leucine we have demonstrated that protein oxidation contributes approximately 8 kcal/kg.d, leaving 20 kcal/kd.d unexplained. This is most likely accounted for by fat oxidation and this has been shown to be an active process during the first days of life, at least in term infants [35]. It can be concluded therefore that the preterm baby, who is using mainly glucose as an energy source before birth, changes rapidly to a combination of glucose and fatty acids to cover energy needs after birth. Whether cold stress changes this pattern is unknown as the necessary studies are virtually impossible to perform. The increasing glycerol levels observed by Heim [32] may indicate that fatty acid oxidation becomes more important during cold stress in the newborn infant.

Conclusion

1. The thermoregulatory response of the fetus is probably inhibited by some (still unknown) mechanism before birth.

2. Animal studies show an important role for brown adipose tissue in non-shivering thermogenesis during the first days after birth, but its role in humans is still unclear. The amount of brown fat in very preterm infants is limited and the quantity of uncoupling protein lower than in term babies.

3. Intracellular conversion of T4 to T3 is important for the activation of brown adipose tissue.

4. Very preterm infants seem incapable of increasing heat production as a response to low environmental temperatures in the first few days of life, while the term baby can respond within the first hours after birth.

5. Defining the most desirable environmental temperature for the very preterm infant during the first days (or weeks) after birth cannot therefore be based on oxygen consumption. Simple methods such as controlling central temperature may be more appropriate.

6. While glucose is the main fuel before birth, lipids become a major energy source soon after birth, even in the absence of cold stress.

References:

1. Budin P (1900) „Le nourisson“, alimentation et hygiène des enfants débiles - enfants nés à terme. Paris Octave Dion ed.

2. Silverman WA, Fertig JW, Berger AP (1958) The influence of the thermal environment upon the survival of newly born premature infants. Pediatrics 22:876-886.

3. Glass L, Silverman WA, Sinclair JC (1968) Effect of the thermal environment on cold resistance and growth of small infants after the first week of life. Pediatrics 41:1033-1046.

4. Silverman WA, Agate Jr. FJ (1964) Variations in cold resistance among small newborn infants. Biol Neonate 6:113-127.

5. Brück K (1961) Temperature regulation in the newborn infant. Biol Neonate 3:65-119.

6. Hey EN (1969) The relation between environmental temperature and oxygen consumption in the newborn baby. J Physiol 200:589.

7. Smales O (1978) Simple method for measuring oxygen consumption in babies. Arch Dis Child 53:53.

8. Casteilla L, Champigny O, Bouilland F, Robelin J, Ricquier D (1989) Sequential changes in the expression of mitochondrial protein mRNA during the development of brown adipose tissue in bovine and ovine species. Biochem J 257:655-671.

9. Polk DH (1988) Thyroid hormone effects on neonatal thermogenesis. Seminars in Perinatol 12:151-156.

10. Thorburn GD, Hopkins PS (1973) Thyroid function in the foetal lamb. In: Comline KS, Cross KW, Dawes GS, Nathalielsz PW (eds) Foetal and Neonatal Physiology. Cambridge University Press, Cambridge, pp 488.

11. Breall JA, Rudolph AM, Heymann MA (1984) Role of thyroid hormone in postnatal and metabolic adjustments. J Clin Invest 73:1418-1424.

12. Polk DH, Wu SY, Fischer DA (1986) Serum Thyroid hormones and tissue 5'-monodeiodinase activity in acutely thyroidectomized newborn lambs. Am J Physiol 251:E151-155.

13. Power GG, Gunn TR, Johnston BM, Nichols G, Gluckman PD (1989) Umbilical cord occlusion but not increased plasma T3 or norepinephrine stimulate brown adipose tissue thermogenesis in the fetal sheep. J Dev Physiol 11:171-177.

14. Gunn TR, Gluckman PD (1989) The endocrine control of the onset of thermogenesis at birth. Baillière's Clin Endocr Metab 3:869-886.

15. Whitaker EM, Hussain SH, Hervey GR, Tobin G, Rayfield KM (1990) Is increased metabolism in rats in the cold mediated by the thyroid? J Physiol 431:543-556.

16. Gluckman PD, Gunn TR, Johnston BM (1983) The effect of cooling in breathing and shivering in unanaesthetized fetal lambs in utero. J Physiol 343:495-506.

17. Gunn TR, Ball KT, Gluckman PD (1991) Reversible umbilical cord occlusion: effect on thermogenesis in utero. Pediatr Res 30:513-517.

18. Gautier H, Bonora M (1992) Ventilatory and metabolic responses to cold and hypoxia in intact and carotid body-denervated rats. J. Appl Physiol 73:847-854.

19. Herpin P, Le Dividich J, Van Os M (1992) Contribution of colostral fat to thermopgenesis and glucose homeostasis in the newborn pig. J Dev Physiol 17:133,141.

20. Santalucia T, Camps M, Castello A, Munoz P, Nuel A, Testar X, Palacin M, Zorzano A (1992) Developmental regulation of GLUT 1 (Erythroid/Hep G2) and GLUT 4 (Muscle/Fat) glucose transporter expression in rat heart, skeletal muscle, and brown adipose tissue. Endocrinology 130:837-846.

21. Castello A, Rodriguez-Manzaneque JC, Camps M, Pérez-Castillo A, Testar X, Palacin M, Santos A, Zorzano A (1994) Perinatal hypothyroidism impairs the normal transition of GLUT 4 and GLUT 1 glucose transporters from fetal to neonatal levels in heart and brown adipose tissue. J Biol Chem 269:5905-5912.

22. Hey En, Katz G (1970) The optimum thermal environment for naked babies. Arch Dis Child 45:328-334.

23. Sauer PJJ, Dane HJ, Visser HKA (1984) New standards for neutral thermal environment of healthy very low birth weight infants in week one of life. Arch Dis Child 59:18-22.

24. Hey E (1974) Physiological control over body temperature. In: Monteith JL, Mount LE (eds) Heat loss from Animals and Man. Butterworth, London.

25. Brück K, Parmelee Jr AH Brück M (1962) Neutral temperature range of „thermal comfort“ in premature infants. Biol Neonate 4:32-51.

26. Wheldon AE, Hull D (1983) Incubation of very immature infants. Arch Dis Child 58:404-408.

27. Lean MEJ, James WPT, Jennings G, Trayhurn P (1986) Brown adipose tissue uncoupling protein content in human infants, children and adults. Clin Sci 71:291-297.

28. Houstek J, Vizek K, Pavelka S, Kopecky J, Krejcova E, Hermanska J, Cermakova M (1993) Type II iodothyronine 5'-deiodinase and uncoupling protein in brown adipose tissue of human newborns. J Clin Endocrinol Metab 77:382-387.

29. Kalhan SC, Denne SC (1990) Energy consumption in infants with bronchopulmonary dysplasia. J Pediatr 116:662-664.

30. Palme-Kilander C, Tunell R, Chiwei Y (1993) Pulmonary gas exchange immediately after birth in spontaneously breathing infants. Arch Dis Child 68:6-10.

31. Sameic TD, Radmacher P, Hill T, Adamkin DH (1994) Measured energy expenditure in mechanically ventilated very low birth weight infants. Am J Med Sci 307:182-184.

32. Heim T (1981) Energy requirements of thermoregulatory heat production in the newly born. In: Monset-Couchard M, Minkowski A (eds) Physiological and biochemical basis for perinatal medicine. S. Karger, Basel, pp 158-174.

33. Glamour TS, McCullough AJ, Sauer PJJ, Kalhan SC (1995) Quantification of carbohydrate oxidation by respiratory gas exchange and isotopic tracers. Am J Physiol 268:E789-796.

34. Van Goudoever JB, Sulkers EJ, Chapman E, Carnielli VP, Efstatopoulos T, Degenhart HJ, Sauer PJJ (1993) Glucose kinetics and glucoregulatory hormone levels in ventilated preterm infants on the first day of life. Pediatr Res 33:583-589.

35. Patel D, Kalhan S (1992) Glycerol metaboliosm and triglyceride-fatty acid cycling in the human newborn: effect of maternal diabetes and intrauterine growth retardation. Pediatr Research 31:52-58.

Chapter 3

Neonatal Heat Transfer, Routes of Heat Loss and Heat Gain

GUNNAR SEDIN

Introduction

Metabolic processes produce heat which is principally distributed in the body by the circulation of the blood but also in part by tissue conduction (for review, see 1). To maintain a constant body temperature the heat production and exchange with the environment must balance. Heat produced by the fetus has to pass through the mother to be eliminated and the umbilical circulation is the major route of heat exchange [2]. Immediately after birth the human infant is exposed to a colder temperature than in utero and at the same time fluid evaporates from the skin, with a resulting loss of heat and a decrease in body temperature. This is partly a physiological response as the body temperature at birth is higher than that in subsequent life. Exposure to cold may give rise to thermogenic responses which will increase basal heat production [3, 4, 5, 6, 7], and the skin circulation may decrease to reduce the heat loss [8]. Undue heat loss in an infant early after birth is usually prevented by wiping and covering the infant's skin. Seriously ill term infants and preterm infants are nursed in an environment in which a normal body temperature can be maintained, either in an incubator at an ambient temperature within the thermoneutral zone, or under a radiant heater.

Heat Transfer and Routes of Heat Exchange

Heat exchange between the infant and its environment occurs through the skin, and to some extent through the respiratory tract, and occurs by conduction (H_{cond}), evaporation (H_{evap}), radiation (H_{rad}) and convection (H_{conv}). The extent of heat transfer depends on the surface area of the infant and the proportions of the body surface in direct contact with the mattress and/or clothing (H_{cond}), losing water through evaporation (H_{evap}), facing surrounding surfaces (H_{rad}) or exposed to the ambient air (H_{conv}). To determine the heat exchange it is therefore necessary to know the heat loss from the skin per unit surface area, the total body surface area and the proportion of the surface area participating in the different modes of heat exchange [9, 10, 11, 12, 13].

Calculation of Heat Exchange Between the Infant's Body Surface and the Environment.

Heat exchange through conduction, evaporation, radiation and convection can be calculated using the following equations [12, 13] and with a knowledge of the transepidermal water loss (TEWL), the temperature of the material on which the infant is placed (Tbed), the temperature of the infant's skin (T_{skin}), the temperature of the ambient air (T_{amb}), the temperature of the walls facing the infant (T_{wall}), and characteristics of the material in the infant's environment :

Heat exchange through conduction:

$$H_{cond} = k_0 \, (T_{skin} - T_{bed}) \qquad (W/m^2)$$

where k_0 represents the conductive heat transfer coefficient. H_{cond} is dependent on the thermal characteristics of the skin, but even more on those of the mattress. T_{skin} (K) is the temperature of the skin and T_{bed} (K) the temperature of the bed (mattress). With the thermal conductivity characteristics of most regular mattresses, the heat loss through conduction in incubators and under radiant heaters is very low.

Heat exchange through evaporation:

$$H_{evap} = k_1 \cdot TEWL \; (3.6 \times 10^3)^{-1} \qquad (W/m^2)$$

where k_1 is the latent heat of evaporation ($2.4 \cdot 10^3$ J/g), TEWL is the transepidermal water loss (g/m^2h) and 3.6×10^3 is the correction factor for time (s). TEWL is a mean value of evaporation of water from the skin surface measured using the gradient method [14, 15, 16].

Heat exchange through radiation:

$$H_{rad} = S_0 \cdot e_1 \cdot e_2 \cdot (T^4{}_1 - T^4{}_2) \qquad (W/m^2)$$

where S_0 is Stefan-Boltzman's constant ($5.7 \cdot 10^{-8} W/m^2K^4$), e_1 is the emissivity of the skin, e_2 is the emissivity of the surrounding walls (0.97), T_1 is the mean temperature of the skin (K) and T_2 is the mean temperature of the surrounding wall (K).

Heat exchange through convection :

$$H_{conv} = k_2 \; (T_1 - T_3) \qquad (W/m^2)$$

where k_2 is the convection coefficient ($2.7 \; W/m^2K$), T_1 is the mean temperature of the skin (K) and T_3 is the mean temperature of the ambient air (K). This calculation does not include fast convections, which in adults occur at air velocities above 0.27 m/s [17].

The extent of the heat exchange between the body surface and the environment depends on the type of heat exchange, on the position and geometry of the body, and on the magnitude and frequency of body movements. Comparisons of heat exchange under different environmental conditions in infants of different gestational and postnatal ages are therefore often presented as heat exchange per unit area of body surface exposed to the ambient air or facing the walls of the incubator.

Heat Exchange Through the Respiratory Tract.

The expired air is usually more humid, i.e. it has a higher water vapour pressure, than the inspired air. This results in an evaporative loss of water and heat from the respiratory tract. A small amount of convective heat transfer also takes place in the respiratory tract and often these two processes are considered together (for review, see 1). In the newborn heat gain through the respiratory tract can also occur.

The alternating displacement of air during the respiratory cycle makes convective and evaporative heat transfer in the respiratory tract complex. When ambient air, which is cooler than the body, passes along the mucosa during inspiration, it gains heat by convection and water vapour by evaporation from the mucosa. When this air reaches the alveoli it is at thermal equilibrium with a central body temperature and is saturated with water. During expiration the air may become a little cooler than the body temperature before it leaves the infant.

Calculation of Heat Exchange from the Respiratory Tract of the Infant.

The exchange of heat through convection in the respiratory tract, $H_{conv\text{-}r}$, is calculated from the air volume ventilated per unit time (V = ventilation volume) and the temperature difference between expired and inspired air (T_E - T_I) according to the following relationship:

$$H_{conv\text{-}r} = V \cdot \rho \cdot c\,(T_E - T_I)\,m^{-1} \qquad \text{(W/kg)}$$

where V is the ventilation volume per unit time, ρ is the density of the air (1 g = 0,880 l), c is the specific heat ($1\ J \cdot g^{-1} \cdot {}^{\circ}C^{-1}$) and m is the body weight (kg) and T_E and T_I the temperatures of expired and inspired air.

As a result of the alternate inspiratory warming and expiratory cooling of the air, the convective heat exchange in the respiratory tract depends mainly on the temperature of the inspired air. In human infants nursed in incubators there is a very small difference between the temperature of the inspired and expired air and convective losses are therefore small.

Evaporative heat exchange through the airway (H_{evap-r}) depends on the difference in water content between expired and inspired air. This is the respiratory water loss [18, 19, 20]. As the formation of water vapour in the respiratory tract requires thermal energy the amount of heat exchange by evaporation per unit time will be:

$$H_{evap-r} = k_1 \cdot RWL \ (3.6 \cdot 10^3)^{-1} \qquad (W/kg)$$

where k_1 is the latent heat of evaporation of water ($2.4 \cdot 10^3$ J/g), RWL is the respiratory water loss (mg/kg min) and $(3.6 \cdot 10^3)^{-1}$ the correction factor for time.

Heat Gain.

More than 100 years ago it was realised that a good thermal environment increased the chances of survival of newborn infants, and this led to the construction of the first incubators (for review, see 21). Budin [22] found increased survival rates in infants whose temperatures had never been below 32 °C. Later studies by Silverman and co-workers [23, 24], Hey and Katz [5], and Dahm and James [25] widened our knowledge of the influence of ambient temperature on the survival rate, oxygen consumption and respiration of newborn infants.

Agate and Silverman [26] introduced the use of infrared radiation to control the body temperature of small newborn infants. Radiant warmers have since been widely used in neonatal intensive care. The contributions of different modes of heat exchange under radiant heaters vary markedly depending on the shields and blankets used [see 27, 28].

Incubators.

In a convectively heated incubator warm, and usually humidified, air is supplied to the hood under which the infant is placed. The warm air is usually directed so that both the air and walls of the incubator are kept warm. At the same time it is desirable to keep the air flow velocity close to the baby less than 0.2 m/s, allowing a normal convective pattern around the infant. At a velocity lower than 0.1 m/s the convective heat transfer depends on the temperature gradient between the skin and the air, and at this low velocity the vapour pressure gradient close to the skin surface is also maintained, avoiding increased evaporative water loss due to air flow velocity [29, 30].

The incubators available today have air flow velocities varying from 0.04 - 0.94 m/s over the bed [31]. Most are also supplied with a system for humidification, in some the ambient relative humidity can be raised to 96 % while others have relative humidities of 60 - 70 % [31]. A high humidity markedly reduces evaporative heat loss [12, 13, 32, 33].

In single-walled incubators 40 % or less of the air space is within 0.5 °C of the set temperature, whereas incubators with double walls have more uniform temperature distributions. The temperature of the incubator walls varies markedly between different types of incubators [31]. Those with single walls generally have cooler inner hood surfaces than those with double walls [31]. There are therefore obvious differences in the contributions of the different modes of heat exchange depending on the incubator in which the baby is nursed.

Radiant warmers.

With radiant warmers the infant is heated only by an overhead panel which produces radiant heat and the heat is transferred from the skin to deeper tissues by conduction and by circulation of blood. Heat received from the heater is sometimes estimated as radiant power density (mW/cm^2) [34]. It is very difficult to estimate the different modes of heat exchange when the infant is nursed under a radiant heater. As there may be free air movements above the body surface of the infant, both insensible water and heat loss and convective heat loss can increase as a result of a high air velocity. In addition, the ambient vapour pressure is usually low in the nurseries, which will increase the evaporative heat losses. The exchange of heat through radiation is an overall gain, but may involve losses due to the temperature of the walls facing the infant sides [28]. The modes of heat exchange between the infant and its environment when nursed under a radiant heater will be presented in a separate chapter.

Relationship Between Different Modes of Heat Exchange

Heat exchange at an ambient humidity of 50 % during the first day after birth:

The calculated exchange of heat between infants of different gestational ages and the environment at an RH_{amb} of 50 % is shown in figure 3.-1 [13]. As the evaporative heat exchange is directly proportional to the amount of water evaporated from the infant's skin, evaporative heat loss shows the same kind of relationship to gestational age as TEWL [13, 16]. In the most preterm infants the evaporative heat exchange may reach 50 W/m^2 of the body surface area, while in term infants it is close to 5 W/m^2.

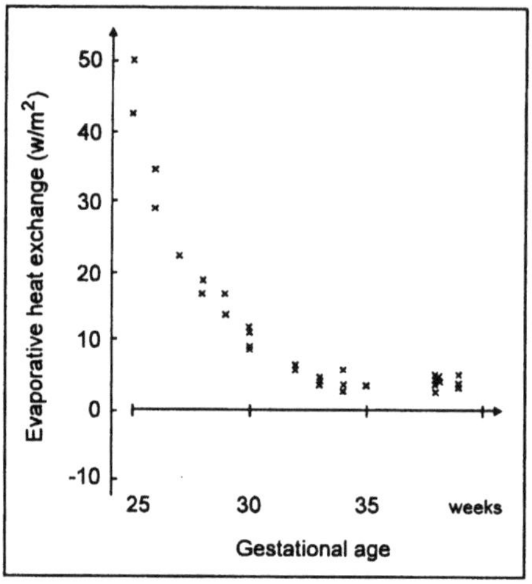

Fig. 3.-1 : Heat exchange trough evaporation in relation to gestational age [13].

Because of the evaporative heat loss in very preterm infants, a very high ambient temperature is needed to maintain a normal body temperature, of between 36.0 °C and 37.0 °C, in these babies. Those born at a gestational age of less than 28 weeks need a T_{amb} of around 40 °C to maintain a normal body temperature at a low humidity of 20 %, while fullterm infants need an ambient temperature of around 34 °C (Figure 3.-2). At a higher ambient humidity a somewhat lower air temperature is needed to maintain a normal body temperature in the most immature infants [13].

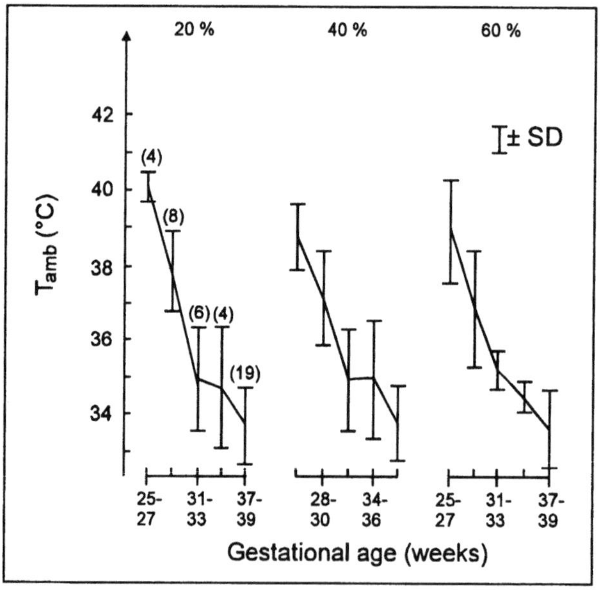

Fig. 3.-2 : Ambient temperature in relation to gestational age at RH_{amb} 20 %, 40 % and 60 %. SD = Standard Deviation [13].

In the very preterm infant the evaporative heat exchange with the environment is highest at a low humidity. At RH_{amb} of 60 % the heat exchange through evaporation in these infants is only 50 % of that at RH_{amb} of 20 % (Figure 3.-3).

The radiative heat exchange is lowest in the most preterm infants and highest in those born at term. This is due to the warmer air in the incubator increasing the temperature of the surfaces facing the infant (Figure 3.-3).

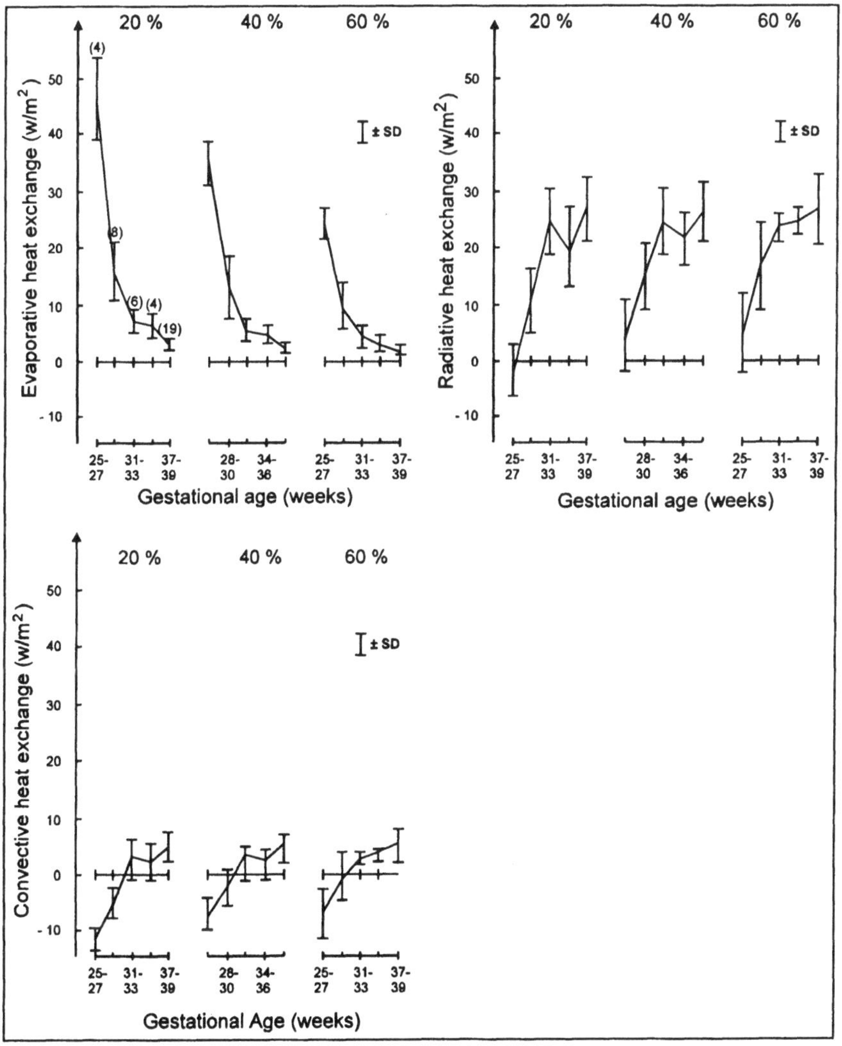

Fig. 3.-3 : Heat exchange through evaporation, radiation and convection in relation to gestational age at RH_{amb} 20 %, 40 % and 60 %. SD = Standard Deviation [13].

The heat exchange by convection is dependent on the gestational age of the infant and on the ambient humidity. At all the humidities studied, heat is gained by convection in infants born between 25 and 30 weeks of gestation, and heat loss by convection is low in moderately preterm and term infants. The differences in heat exchange between infants of different gestational ages are due to the gestational age dependency of the evaporation rate from the skin, and thus the rate of evaporative heat loss, figure 3.-1. To keep the infant's thermal balance at a steady state, the servo-control system of the incubator, which controls the skin temperature, increases the temperature of the air, which also leads to an increase in the temperature of the incubator walls. Thus, the loss of heat to the environment through convection and radiation decreases, and most preterm infants even gain heat through these modes of heat exchange. As these data were obtained while the infants were being nursed in an incubator with a relative humidity of 50 %, the preterm infants who needed a higher ambient temperature were nursed at a higher ambient vapour pressure than more mature infants. The evaporative heat exchange would have been even higher if the ambient vapour pressure had been kept constant instead of RH_{amb}.

If the different modes of heat exchange from the skin surface are added together the sum will be fairly constant throughout the gestation age groups. The total heat loss cannot however be calculated in this way as the proportions of the body surface area exchanging heat in the different ways are not precisely known.

Heat Exchange During the First Weeks After Birth.

During the first weeks after birth the temperature of the incubator air can gradually be lowered both in the preterm and the term infant [32]. The high evaporative heat losses in the first days after birth in the most preterm infants will gradually decrease with increasing postnatal age (Figure 3.-4) [35]. Heat loss through radiation, which is initially low just after birth in the most preterm infants born at 25 - 27 weeks of gestation, will be the most important mode of heat exchange after the first postnatal week. In infants born at a gestation of more than 28 weeks the radiative heat exchange is the most important mode of heat exchange from birth. The heat loss through radiation will gradually increase with age. The most preterm infants will gain heat by convection over the first 10 days after which this mode will result in a small heat loss.

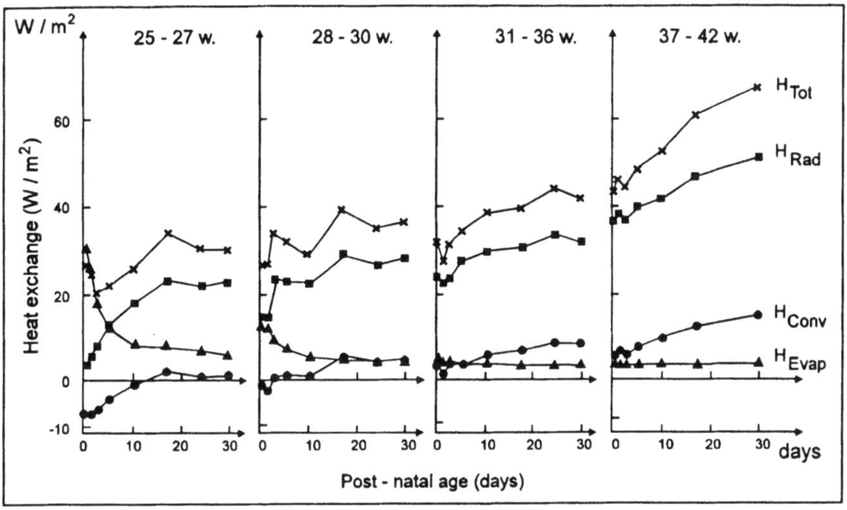

Fig. 3.-4 : Heat exchange between the infant and the environment per m^2 BSA in relation to postnatal age in different gestational age group at an ambient humidity of 50 % [32].

During the first weeks after birth the relationship between the different modes of heat exchange depends on the ambient humidity (Figure 3.-5). In a dry environment, in infants born at 25 - 27 weeks gestation, evaporative heat exchange will be the most important mode for more than 10 days, while at an ambient humidity of 60 % this mode of exchange will be much lower and will be exceeded by the heat exchange through radiation from the 5th day after birth.

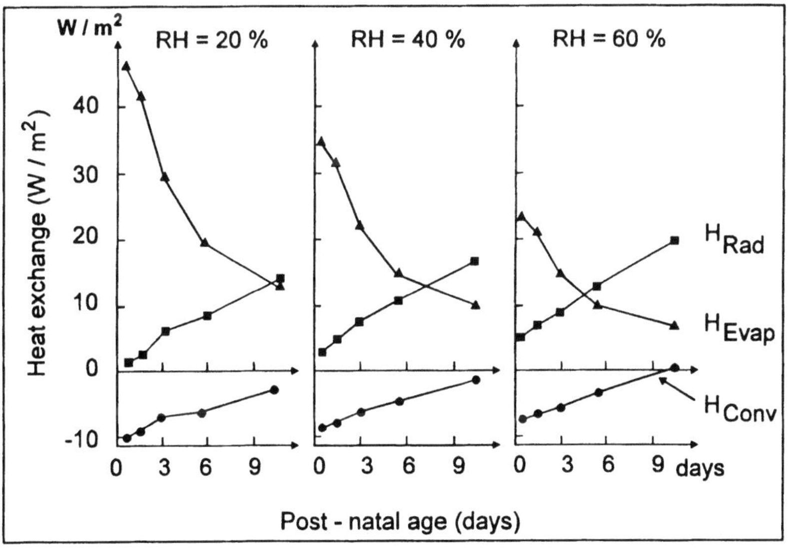

Fig. 3.-5 : Heat exchange between the infant and the environment in relation to postnatal age at ambient humidity 20 %, 40 % and 60 % in infants born at 25 - 27 week of gestation [32].

The Total Heat Exchange between the Infant's Skin and the Environment.

To calculate the total heat exchange between the infant's skin and environment, the body surface area of the infant and the fractions of this area that participate in each mode of heat exchange have to be calculated. With available methods it is very difficult to determine how different fractions of heat exchange vary over a longer period of time. In figure 3.-6 the upper curve represents the total heat exchange when the whole body surface area participates in all modes of heat exchange. The lower curve for each group of infants was calculated under the assumption that smaller proportions of the body surface area exchange heat by evaporation, convection and radiation. The changes with age will be almost the same for all these curves.

Heat Loss from the Respiratory Tract.

The evaporative loss of heat from the respiratory tract can be determined using the measured values of respiratory water loss. As the respiratory water loss in fullterm infants is of about the same magnitude as their insensible water loss from the skin, if the infants are nursed in an ambient humidity of 50 %, the evaporative heat loss from the airway will be moderate [19]. Data from newborn lambs [18] has shown that the evaporative loss of water and heat from the airway depends on the ambient humidity and on the postnatal age. The respiratory water loss is directly proportional to the rate of breathing [36, 37], and infants therefore lose more water and heat when they are tachypnoeic.

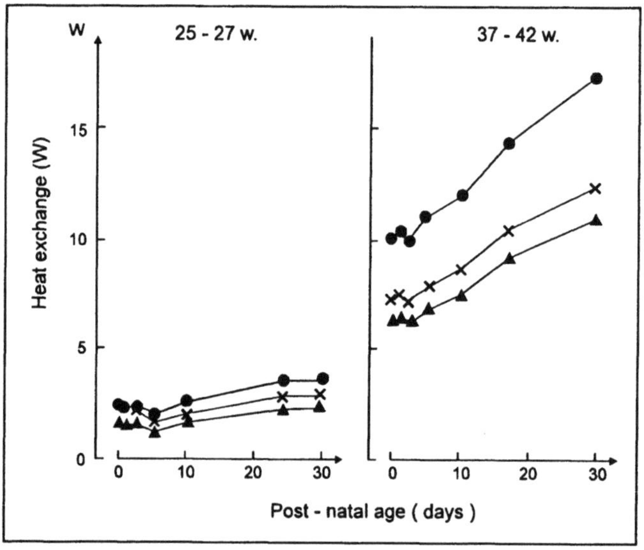

Fig. 3.-6 : Heat exchange between the infant and the environment on participation of different body surface area fractions in each mode of heat exchange, in relation to postnatal age in infants born after 25-27 weeks of gestation and in fullterm infants. The curves are based on the following fractions: Upper curve: H_{evap} 1.0, H_{rad} 1.0, H_{conv} 1.0; Middle curve: H_{evap} 1.0, H_{rad} 0.7, H_{conv} 0.7; Lower curve: H_{evap} 0.7, H_{rad} 0.6, H_{conv} 0.7 [32].

Infants can tolerate moderate heat stress without increasing their respiratory water loss [38], but with increasing activity [20] the respiratory water and heat loss may increase up to 140 % in term infants.

Infants who inspire air that is warmer than the temperature of the airway mucosa will gain heat through convection, and those inspiring air cooler than the airway mucosa will lose heat through this mode of heat exchange. Infants who inspire cold, dry air will have a high loss of heat from the respiratory tract.

Summary and Conclusion

The exchange of heat between the infant's skin and the environment is influenced by the metabolic rate, the insulation and permeability of the skin as well as environmental factors such as the ambient temperature and humidity, air flow velocity and the temperature and characteristics of incubator surfaces facing the infant. Evaporative heat loss from the skin is the major component of heat exchange in the most preterm infants soon after birth and these babies gain heat through convection and, in a very dry environment, possibly also through radiation when they are nursed in incubators. As the water loss from the skin surface of these most preterm infants decreases with postnatal age, the heat loss through evaporation also decreases. Their need for a high ambient temperature also diminishes and with the lower temperature of the incubator walls, the heat loss through radiation will then increase and the heat gain by convection will be changed to a low loss of heat during the first weeks after birth. In infants nursed under a radiant heater, heat is gained through radiation, and evaporative losses of water and heat may be very high in preterm infants. The loss of heat through convection depends on how the infant is protected from high air velocities while under the radiant heater, and also on the magnitude of air movement within the nursery.

In the term baby, the respiratory water loss at an ambient humidity of 50 % is of the same magnitude as the insensible water loss from the skin, and both therefore result in a small heat loss in these infants. In the preterm evaporative heat losses from the skin are of much more importance than such heat losses from the respiratory tract. The evaporative heat loss from the airway might be considered important only in infants with high motor activity and with tachypnoea, especially if they are nursed in a very dry environment. Ventilation with dry and cold air will also result in a high loss of heat through evaporation and convection.

Implications for Practice

- Wipe and cover the newborn infant with a warm towel
- Place seriously ill infants and preterm infants in an incubator, prewarmed to appropriate temperature
- Use a high ambient humidity in the incubator to lower heat loss through evaporation in preterm infants
- Avoid frequent opening of ports and when necessary prevent leakage of warm air out of the incubator
- Heat and humidify all gas delivered by the ventilator to keep heat loss from the airway low

Recommendation for Future Research

Develop an incubator that allows the user to set the wall temperatures, air temperature, mattress temperature and vapour pressure of incubator air and airflow velocity independently, and study the influence of changing these variables one by one while keeping the others constant.

Abbreviations

H_{cond},	heat exchange through conduction between the skin and the environment (W/m^2)
H_{evap},	heat exchange through evaporation from the skin surface (W/m^2)
H_{rad},	heat exchange through radiation between the skin and environment (W/m^2)
H_{conv},	heat exchange through convection between the skin and the environment (W/m^2)
H_{evap-r},	heat exchange through evaporation from the respiratory tract
H_{conv-r},	heat exchange through convection in the respiratory tract
P_{H2Oamb},	water vapour pressure of ambient air
RH_{amb},	relative humidity of ambient air
TEWL,	transepidermal water loss (g/m^2 x h)
T_{bed},	temperature of bedding (mattress) (K)
T_{skin},	skin temperature (K)
T_{amb},	temperature of ambient air (K)
T_{wall},	temperature of the wall of an incubator or a room (K)
T_E,	temperature of expired gas (K)
T_I,	temperature of inspired gas (K)
ρ,	density of the ventilation gas
V,	ventilation volume (l/unit time)
c,	specific heat of gas
m,	body mass (kg)

Acknowledgement. This paper is based on studies supported by the Swedish Medical Research Council (project 19X-04998).

References:

1. Houdas Y, Ring EFJ (1982) Human body temperature. Its measurement and regulation. Plenum press, New York and London

2. Brück K (1978) Heat procuction and temperature regulation. In Perinatal Physiology, ed. Stave. Publisist 21:455-521

3. Gandy GM, Adamsons K Jr, Cunningham N, Silverman WA, James LS (1964) Thermal environments and acid-base homeostasis in human infants during the first few hours of life. J Clin Invest 43:751-758

4. Brück K (1961) Temperature regulation in the newborn infant. Biol Neonate 3:65

5. Hey EN, Katz G (1969) Evaporative water loss in the newborn baby. J Physiol 200:605-619.

6. Tunell R (1975) The influence of different environmental temperatures on pulmonary gas exchange and blood gas changes after birth. Acta Paediatr Scand 64:57

7. Smales ORC, Hull D (1976) Metabolic response to cold in the newborn. Arch Dis Child 53:407-410

8. Sjörs G, Hammarlund K, Öberg PÅ, Sedin G (1992) An evaluation of environment and climate control in seven infant incubators. Biomedical instrumentation & technology, 294-301

9. Hey EN, Mount LE (1967) Heat losses from babies in incubators. Arch Dis Child 42:75-84

10. Colin J, Houdas Y (1967) Experimental determination of coefficient of heat exchanges by convection of human body. J Appl Physiol 22:31-38

11. Swyer PR (1978) Heat loss after birth. In Sinclair J C (ed): Temperature regulation in energy metabolism in the newborn. Grune & Stratton

12. Hammarlund K, Nilsson GE, Öberg PÅ, Sedin G (1980) Transepidermal water loss in newborn infants. V. Evaporation from the skin and heat exchange during the first hours of life. Acta Paediatr Scand 69:385-392

13. Hammarlund K, Sedin G (1982) Transepidermal water loss in newborn infants. VI. Heat exchange with the environment in relation to gestational age. Acta Paediatr Scand 71:191-196

14. Nilssons G (1977) Measurement of water exchange through skin. Med Biol Eng Comput 15:209-218

15. Hammarlund K, Nilsson G, Öberg PÅ, Sedin G (1977) Transepidermal water loss in newborn infants. I. Relation to ambient humidity and site of measurement and estimation of total transepidermal water loss. Acta Paediatr Scand 66:553-562

16. Hammarlund K, Sedin G (1979) Transepidermal water loss in newborn infants. III. Relation to gestational age. Acta Paediatr Scand 68:795-801

17. Cooney DO (1976) Biomedical engineering principles. An introduction to fluid, heat, and mass transport processes. New York and Basel: Marcel Dekker Inc., 93-155

18. Hammarlund K, Riesenfeld T, Sedin G 1(986) Measurement of respiratory water loss in newborn lambs. Acta Physiol Scand 127:61-65

19. Riesenfeld T, Hammarlund K, Sedin G (1987) Respiratory water loss in fullterm infants on their first day after birth. Acta Paediatr Scand 76:647-653

20. Riesenfeld T, Hammarlund K, Sedin G (1987) Respiratory water loss in relation to activity in fullterm infants on their first day after birth. Acta Pediatr Scand 76:889-893

21. Glatzl-Hawlik M-A, Bell EF (1992) Environmental temperature control In Poline RA, Fox WW (eds) Fetal and Neonatal Physiology, vol 1, W B Saunders Company, 515-526

22. Budin P (1900) Le Nourrison. Paris: Dion

23. Silverman WA, Fertig JW, Berger AP (1958) The influence of the thermal environment upon the survival of the newly born premature infant. Pediatr 22:876-886

24. Silverman WA, Sinclair JC, Agate FJ Jr (1966) The oxygen cost of minor changes in heat balance of small newborn infants. Acta Pediatr Scand 55:294-300

25. Dahm LS, James LS (1972) Newborn temperature and calculated heat loss in the delivery room. Pediatr 49:504-513

26. Agate FJ, Silverman WA (1963) The control of body temperature in the small newborn infant by low-energy infra-red radiation. Pediatr 31:725-733

27. Doyle LW, Sinclair JC (1982) Insensible water loss in newborn infants. In Clinics in Perinatology. Symposium on Fluid Balance in the Newborn Infant, WB Saunders Company 453-482

28. Baumgart S (1982) Radiant energy and insensible water loss in the premature newborn infant nursed under a radiant warmer. In Clinics in Perinatology. Fluids Balance in the Newborn Infant, W B Saunders Company 483-503

29. Okken A, Blijham C, Franz W, Bohn E (1982) Effects of forced convection of heated air on insensible water loss and heat loss in preterm infants in incubators. J Pediatr 101:108-112

30. Thompson MH, Stothers JK, McLellan NJ (1984) Weight and water loss in the neonate in natural and forced convection. Arch Dis Child 59:951-956

31. Sjörs G, Hammarlund K, Kjartansson S, Riesenfeld T, Sedin G (1994) Respiratory water loss and oxygen consumption in fullterm infants exposed to cold air on the first day after birth. Acta Pediatr 83:802-807

32. Hammarlund K, Strömberg B, Sedin G (1986) Heat loss from the skin of preterm and fullterm newborn infants during the first weeks after birth. Biology of the Neonate 50:1-10

33. Sauer PJJ et al (1984) Influence of variations in ambient humidity on insensible water loss and thermoneutral environment of low birth weight infants. Acta Paediatr Scand 73:615-619

34. Baumgart S, Engle WD, Fox WW, Polin RA (1981) Radiant warmer power and body size as determinants of insensible water loss in the critically ill neonate. Paediatr Res 15:1495-1499

35. Baumgart S, Fox WW, Polin RA (1982) Physiology implications of two different heat shields for infants under radiant warmers. J Pediatr 100:787-790

36. Riesenfeld T, Hammarlund K, Jonzon A, Sedin G (1988) Influence of radiant heat stress on respiratory water loss in new-born lambs. Biology of the Neonate 53:290-294

37. Riesenfeld T, Hammarlund K, Norsted T, Sedin G (1994) The temperature of inspired air influences respiratory water loss in young lambs. Biology of the Neonate 65:326-330.

38. Riesenfeld T, Hammarlund K, Sedin G (1990) The effect of a warm environment on respiratory water loss in fullterm newborn infants on their first day after birth. Acta Pediatrica Scandinavica 79:893-898

Chapter 4

Body Temperature in Sick Neonates, Diseases and Biochemical Disturbances

JEAN L. MICHELI

Over the last 30 years, in parallel with the improvements in neonatal intensive care, there has been a gradual shift in the neonatologist's everyday's concerns about dysregulation of the body temperature in high risk newborns.

Nowadays, beside the classic problems [1, 2] including those related to the cold injury syndrome [3], to the post-asphyxia thermal disorders [4] and to increased body temperature in cyanotic heart disease, the part of efforts devoted towards understanding thermic imbalance situations in very low birthweight preterm infants becomes increasingly important.

The following situations of thermal imbalance, either because they are new, or frequent in very low birhtweight preterm infants, or because of their biochemical and metabolic backgrounds, deserve to be discussed here.

Thermal Balance Changes During High Frequency Oscillatory Ventilation (HFO)

It is more than 20 years since HFO was first described [5]. In the intervening period HFO had ups and downs and randomized prospective studies gave conflicting results [6-9]. While waiting for the perfectly designed and controlled study to give a definitive answer whether or not HFO represents an advance in respiratory management, many neonatal intensive care units started gaining clinical experience with HFO. Intriguingly, those very low birthweight preterm baby who seem to respond the best to HFO, tend to decrease their body temperature. This thermal imbalance is easily corrected by an increase of ambient temperature.

From the purely physical stand-point, the way in which gas transport (including water vapour) takes place along the airways, is becoming fairly well understood. The highest rates of gas transport are obtained when the pressure timecourse of oscillation approaches a true sinusoidal waveform [10]. Any condensation of water in the tubing will alter the waveform and decrease the efficiency of HFO [10].

At 37 °C and PH_2O 47 mm Hg, HFO pressure swings (15 Hz; peak to peak pressure 50 -70 cm H_2O) will generate condensation, whereas conventional ventilation will not (~1 Hz; peak inspiratory to end expiratory pressure difference 20-25 cm H_2O). Thus when using HFO ventilation, the neonatologist has a choice between two unsatisfactory alternatives: (a) to keep inspiratory water pressure high and accept suboptimal O_2 and CO_2 gas exchange, (b) to seek for optimal O_2 and CO_2 exchanges at the expense of a lower inspiratory water pressure.

In case the latter alternative has to be choosen, it is possible to run the ventilator's humidifier at 32.5 - 34.5 °C rather than the usual 37 °C for conventional ventilation. The resulting respiratory water and heat losses can be estimated respectively between 8 - 15 ml H_2O/kgxd and 4.7 - 8 kcal/kgxd. This represents roughly 15 % - 20 % of the metabolic heat production. The thermal imbalance [11, 12], needs to be corrected by increasing the incubator's temperature by roughly 1 °C (fig. 4.-1 and 4.-2).

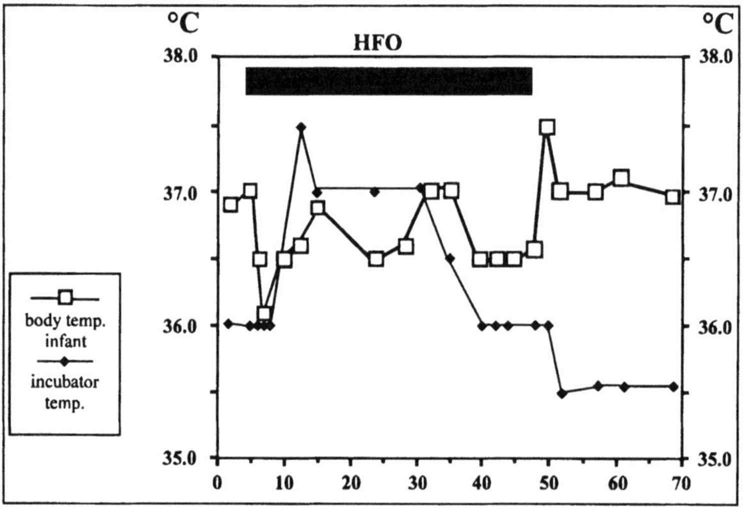

Fig. 4.-1 :Postnatal time-course of body temperature and incubator operating temperature in a preterm infant (30 weeks, birth weight 1080 g), with severe hyaline membrane disease (a/A ratio = 0.12 at a postnatal age of 4h). The initial setting of the incubator's operating temperature corresponds to Sauer's nomogram [33], and the infant' body temperature was 36.9 - 37.0 °C. Soon after starting HFO ventilation, the infant's temperature dropped from 37.0 to 36.0 °C and the incubator's T° had to be increased to 37.5 °C. During the following 24 h the incubator's T° had to stay above the body T° in order to maintain the infant's thermal balance. At the end of the HFO ventilation, the body T°increased sharply.

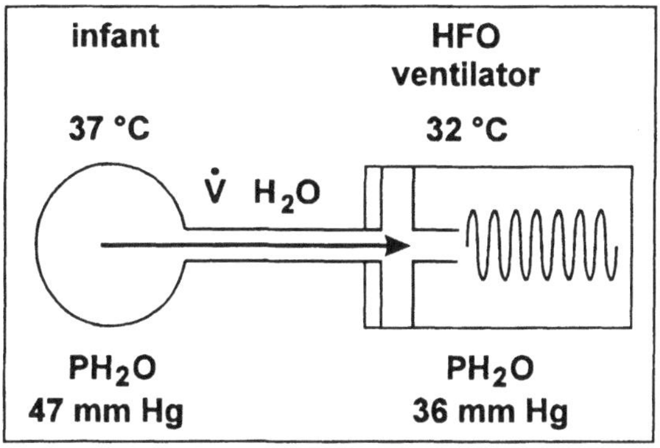

Fig. 4.-2 : Respiratory water losses during HFO ventilation. One of the critical factors for facilitated diffusion of O_2 and CO_2, is the sinusoidal waveform of the oscillation. Any condensation of water in the tubing will alter the waveform and reduce the efficiency of gas exchanges. If HFO ventilators are operated at a high humidification (PH_2O 47 mm Hg, corresponding to 100% saturation at 37 °C), the pressure swings (peak-to-peak 50-70 cm H_2O for example) will facilitate condensation of water and the exchanges O_2 and CO_2 will be suboptimal [10]. If HFO ventilators are operated at a lower humidification (PH_2O 36 mm Hg, corresponding to 100 % saturation at 32 °C), the pressure swings keep their sinusoidal waveform and the exchanges conditions for O_2 and CO_2 will be improved. The respiratory water and heat losses can be estimated to be respectively around 8-15 ml H_2O/kgxday and 4.7-8 kcal/kgxday. The resulting thermal imbalance can be corrected by increasing the incubator's temperature of 1 °C above those obtained via Sauer's nomogram [33].

Inversion of Temperature Gradients and Skin Evaporative Heat Loss

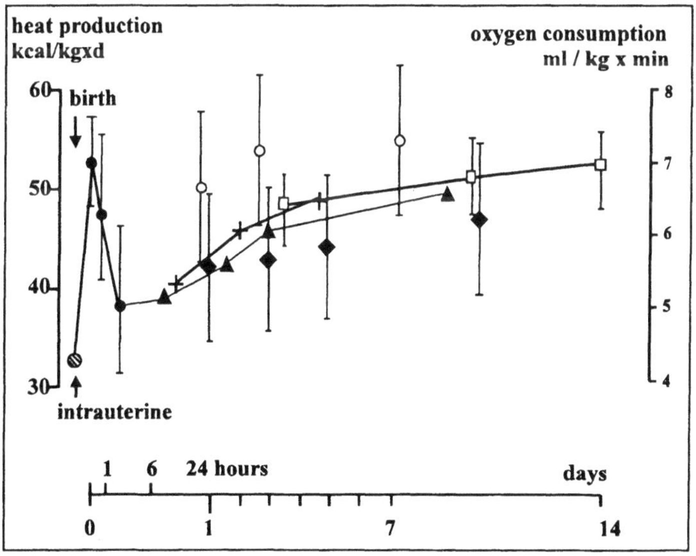

Fig. 4.-3 : Postnatal timecourse of oxygen consumption and energy expenditure. A fair approximation of energy expenditure can be obtained by multiplying oxygen consumption by the caloric equivalent of 1 litre oxygen (4.8 kcal per litre O_2). Adapted from [1, 2, 12, 34].

The postnatal timecourse of oxygen consumption and heat production has been quantified by direct and indirect calorimetry, as well in term, as in preterm newborns [13-15]. During the first 24 - 48 h, oxygen consumption and heat production represent respectively 5 to 7 ml O_2/kgxmin and 35 to 50 kcal/kgxday (fig. 4.-3). A fair approximation of heat production can be obtained by multiplying oxygen consumption by the caloric equivalent of 1 litre of oxygen (4.8 kcal per litre O_2).

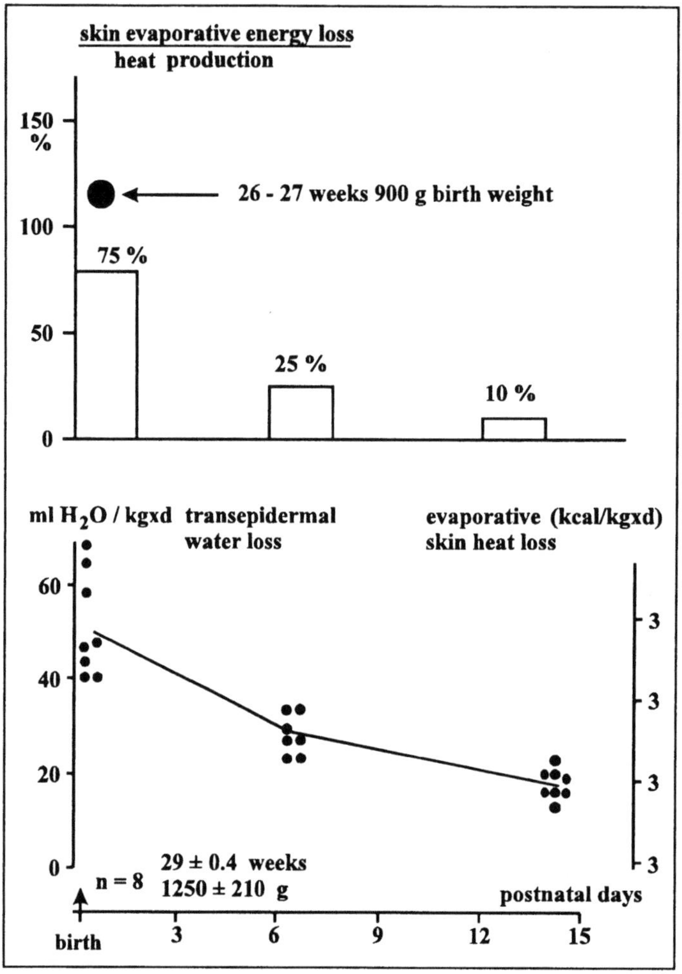

Fig. 4.-4 : Postnatal timecourse of transepidermal water loss in a group of preterm infants (n=8, 29±0.4 gestational weeks, birthweight 1250±210g) nursed in thermoneutral conditions with 50-70 % relative ambient humidity. The corresponding evaporative heat loss (in kcal/kgxday) are shown on they-axis at the right of the diagram.
The upper part of the diagram shows how much of the heat production (measured via indirect calorimetry) is spent to cover the evaporative skin heat losses. Adapted from [17, 18, 34]. In very preterm infants evaporative heat skin losses can exceed the heat produced by the infant's oxidative metabolism. Survival is possible if the ambient temperature is higher than the infant's body temperature.

The lack of cornified epidermis results, even at high ambient relative humidity (50-70 %), in evaporative losses from the skin as high as 100 ml/kgxday for preterm infants <27 weeks during the first 2 postnatal days [16-21]. Each ml of evaporated water represents a deficit of 0.58 kcal. Thus, very preterm infants loose more heat by evaporation [16] than they produce by their oxidative metabolism (fig. 4.-4). Survival is only possible, if the temperature gradient is inversed (ambient T°> body T°) and heat is transferred from the environment into the infant.

Intracellular Energy Failure, Thermal Balance and Glucose Intolerance

Body temperature results from the balance between heat production and heat loss. Since the end of the 1950's, considerable attention has been paid to the increase in oxygen consumption resulting from the exposition to a cold environment and to the catecholamine-mediated nonshivering thermogenesis in brown fat. The reduction of heat losses with extra warmth tripled the chances of survival in very small newborns [22, 23].

Much less attention has been given to the possible consequences of a decreased heat production consecutive to a decreased energy metabolism.

The advent of the phosphorus magnetic resonance spectroscopy of the neonatal brain and the possibility to measure intracellular ATP in vivo, introduced the concept of intracellular energy failure [24, 25]. This condition, which not only occurs in the brain but also in all metabolically active cells, results from a lack of oxygen and/or from a lack in substrate.

The major energy substrate in very low birthweight preterm infants is glucose [26], provided glucose enters into the cells.

The understanding of the entry of glucose into the cells and it's link to the level of circulating amino-acids improved in parallel with the progresses in molecular biology. In that respect, a particularly interesting group of molecules is the family of glucose transporters. Their molecular physiology as well as their regulation have recently been fairly well established [27, 28]. Briefly, a shortage in amino acid can rapidly : (a) reduce insulin and insulin-like growth factors, (b) down regulate glucose transporters at the cellular membrane level and create a glucose intolerance, (c) result in intracellular energy failure.

Hyperglycemia is a common problem in very low birthweight preterm infants with reported incidences as high as 20 % to 86 % [29] particularly during the first few days of life. It is often associated with non oliguric hyperkalemia (fig. 4.-5). The risks linked to this problem are partly due to hyperosmolality, glucosuria and dehydratation; the major risks, however, are intracellular energy failure and Na+/K+ ATP ase deficiency [30].

As far as body temperature is concerned, intracellular energy failure can remain completely unnoticed, because a minimal change in the incubator setting will compensate the decreased heat production. This manoeuvre however, very unlikely is going to compensate a lack of ATP.

As far as growth and development are concerned, the long term consequences of an early postnatal intracellular energy failure should not be minimized [31].

There is a common practice in intensive care units to start i.v. supplies with a simple glucose infusion and refrain from giving iv. amino acid during the first postnatal day. Amino-acids are added to the second day's perfusion, if needed. This traditional way of doing, apparently has no major drawbacks in the majority of neonates except for the very preterm. In these extreme cases (<27 gestational weeks, <1000 g birthweight), high blood glucose and potassium values can be observed, even at glucose infusion rates below placental glucose transfer rates. This clinical observation suggests that, if a minimal amount of amino-acids is not supplied from the first day of life on, glucose intolerance can ensue (fig. 4.-5).

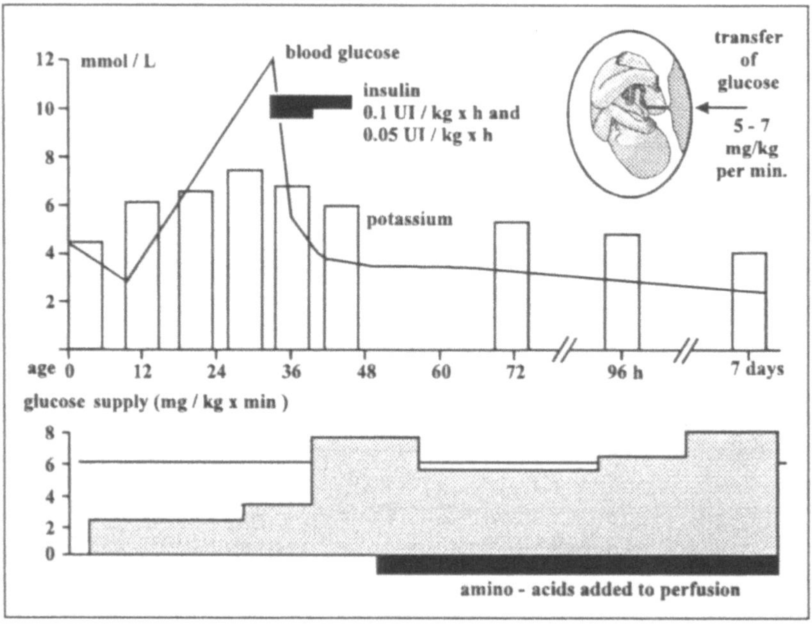

Fig. 4.-5 : Postnatal timecourse of serum glucose and potassium in a very low birthweight preterm infant (26 weeks, birth weight 810 g). During the first 48 hours, the only nutritional support was a 10 % i.v. glucose infusion; i.v. amino acids were added thereafter. The delay in the supply of amino acids, can explain the following serious metabolic imbalances: In spite of a glucose supply (lower part of the diagram) subtantially lower than the expected placental transfer at the same gestational age (visualized by an interrupted line at the 6 mg/kg per min. level), blood glucose and potassium rose. This situation not only reflected a glucose intolerance but also an energy deficit at the cellular level with a shift of potassium ions from the intracellular to the extracellular space [7]. This metabolic imbalance was reversed with a short insulin course. No further insulin was needed after introduction of i.v. amino-acids.

Thermogenic Effect of Feeding, (under-) Nutrition and Body Temperature.

There is a conceptual problem that stems from the fact that the thermogenic response to a meal has been investigated in adult human physiology, where premeal values were considered to reflect a fasting state. In preterm infants this is never the case. With present day's nursing routines (6 - 12 meals/day), the premeal baseline cannot be regarded as a basal postabsorbtive value.

"Healthy" prematures are constantly under food stimulation and part of their thermogenic responses are still present in the following premeal energy expenditure. Therefore the greater the number of meals per day, the smaller the expected postprandial response. At 6-8 meals per day, the thermogenic response to food vary between 3 and 3.5 kcal/kgxday (fig. 4.-6).

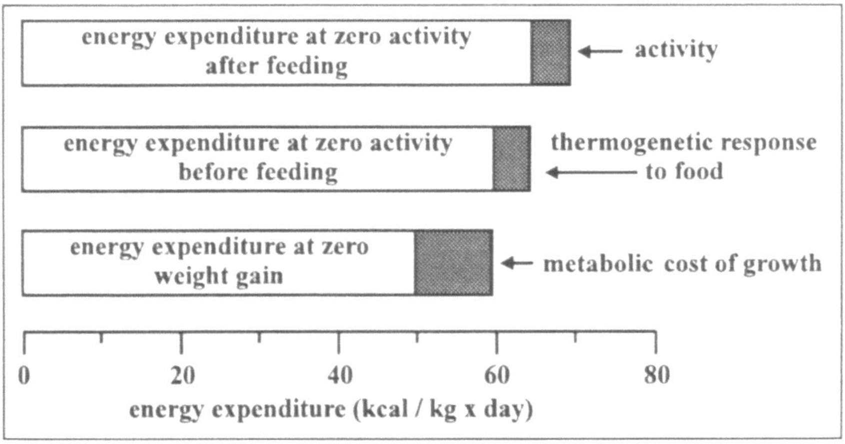

Fig. 4.-6 : Partition of energy expenditure between activity, thermogenic response to food and growth measured in 9 appropriate for gestation age preterm infants (33 ± 1.2 weeks; 1740±180 g birthweight, growth rate at time of study 17 g/kgxday). The upper bar represents the overall energy expenditure over 24 h (68 kcal/kgxday). The bar in the middle allows the comparison between pre- (62 kcal/kgxday) and postmeal (64,5 kcal/kgxday) energy expenditure at the zero activity level: the difference in energy expenditure represents the thermogenic effect of food. The lower bar shows that the energy expenditure related to physical activity and thermogenic effect of feeding is relatively small compared to the metabolic cost of growth (8.5 kcal /kgxday). Adapted from [14, 35, 36].

Sick preterm infants often do not tolerate full feeding, thus only suboptimal amounts of calories and protein are supplied. In these conditions of (under-) nutrition all growth processes, including protein turnover are slowed down and both, the thermogenic response to food and the heat produced by the metabolism of growth are decreased. This could result in hypothermia, if the ambient temperature is not increased to match the sick preterm infant's needs [33].

The opposite can be observed in small for gestational age infants during catch-up growth. They grow faster, their oxygen consumption and heat production are higher, and their range of thermoneutral environmental temperature is lower when compared to their appropriate for gestational age counterparts [34 -36].

Circulatory Problems, Body Temperature and Catecholamines.

The abrupt transition from intra-uterine to extra-uterine life at birth is associated to new environmental stimuli and new demands are made on regulatory mechanisms. For example, the surge in catecholamines observed in term babies at birth, is needed to enhance heat production in response to a relatively cold environment, and to increase myocardial performance and blood pressure during the phase of postnatal circulatory changes [37-40].

In preterm infants, the mass of their adrenal glands, expressed per kg weight, is substantially lower when compared to term infants, and places them in a situation of relative adrenal insufficiency [40]. Their cord blood catecholamine levels are significantly lower [41], and their ability to withstand a cold stress is decreased [13].

The early postnatal timecourse of cardiovascular vital signs observed in "healthy" very preterm infants (<27 gestational weeks, <1000 g birthweight; without hypothermia, hypoxia, acidosis nor infection on admission), reflects this relative insufficiency in catecholamine secretion (Figure 4.-7). During the first 6 to 12 postnatal hours, even though periphery seems adequately perfused, a gradual drop in heart rate and mean arterial blood pressure can be observed. When heart rate drops to values below those of a fetus of the same gestational age (140-150/min), mean arterial blood pressure crosses the lower limits of normal values (25-28 mm Hg). If not treated, this condition could results in a decrease in cardiac output.

The circulatory condition can be improved, either by an increase in fluid (and volume) supply, or by adding a continuous perfusion of catecholamines. The former is simpler, but it bears the risk of overhydratation [38]. The latter could be the more physiologic, because preterm infants are shorter on catecholamines than they are on plasma volume during their first hours of life.

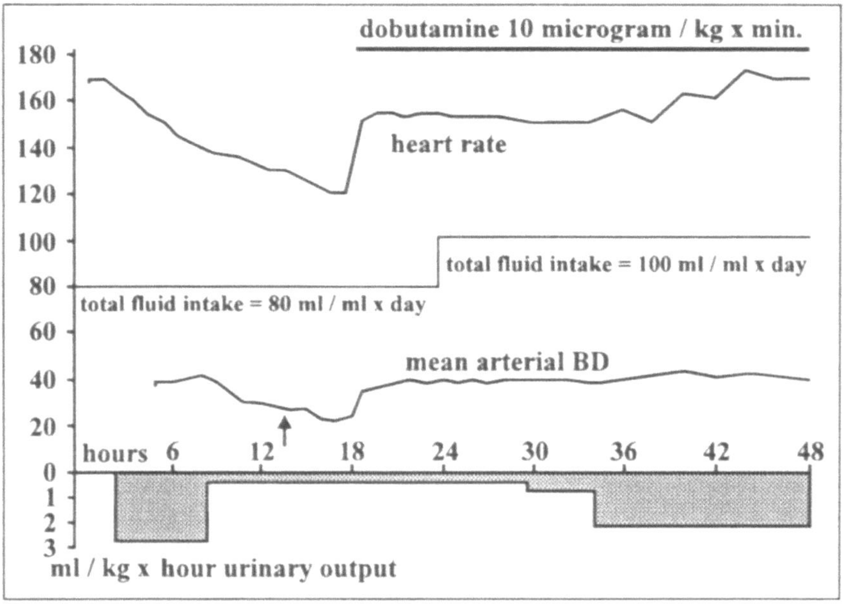

Fig. 4.-7 : Heart rate and mean arterial blood pressure in a "healthy" very low birthweight preterm infant (26 weeks, birth weight 850 g) without asphyxia nor infectious problems and a minimal need of ventilatory support. In the hours following birth both, heart rate and blood pressure decreased in parallel. By the time heart rate fell below mean intrauterine values for a fetus of 26 weeks, blood pressure reached the lower normal range [42] and urinary output became low. Current practice, in this situation, is either to increase fluid (and volume) supply, or to administer catecholamines. The former attitude seems more simple (but bears the risk of overhydratation), the latter may be more physiologic because very low birthweight preterm infants are shorter on catecholamines than they are on plasma volume and total body water [43].

"Therapeutic" Hypothermia.

Intentional hypothermia is successfully used with cardiopulmonary bypass in infant cardiac surgery. A few uncontrolled studies have reported on the use of hypothermia for the treatment of perinatal asphyxia. The rationale for inducing hypothermia as an adjunct to treatment depends on a reduction of rate of oxygen consumption at very low body temperature. In a clinical condition in which oxygen supply is limited, hypothermia could be beneficial if there were a more favourable balance between oxygen uptake and oxygen requirement at low rather than at normal body temperature [1, 4]. On the basis of these uncontrolled reports, it is not possible to make any recommendations regarding hypothermia for the resuscitation of asphyxiated infants.

PGE2, Fever and Respiratory Control.

When confronted by an infant with a possible cyanotic cardiac disease, there is an indication to begin an i.v. infusion of prostaglandin E1 (PGE1) at 0.05 microgram/kgxmin., while further evaluation is undertaken. In general, the more severe the cyanosis, the more urgent the administration of PGE1, because a ductal closure in such newborns would be life threatening. The principal difficulty with PGE1 administration is the occurrence of apnea and fever [32]. From the physiologic point of view, this observation is very interesting, because it shows the close link between the regulation of temperature and the regulation of breathing. From the clinical point of view it is reasonable to intubate all infants receiving PGE1, because apnea may not be immediate, and to monitor vital signs including body temperature and ambient temperature. Prostaglandins E have a pyrogenic action, they produce a set point shift by interacting with thermosensitive structure in the hypothalamus and the mechanisms of fever (peripheral vasoconstriction, thermogenesis) are activated [1]. If this rise in temperature is mistakenly thought to be due to overheating, and the ambient temperature is lowered, there is a risk of worsening the already serious clinical condition.

However, if the ambient temperature is kept unchanged and the infant's body temperature allowed to raise up to 38.5 °C, the PGE1 infusion can be followed by a dramatic improvement.

Implication for Practice

1. In spite of major progresses in neonatal care, thermal imbalances and dysregulation of the body temperature remain a matter of concern for every practitioner.

   In view of the last decade's new therapeutic approaches to the increasing number of very low birthweight preterm infants, the following wishes can be addressed :

2. To have all the tubings including the humidifier of HFO ventilators within the incubator and at slightly a warmer temperature (1-2 °C) than the operating temperature of the incubator, but at a PH_2O of 47 mmHg. This in order to prevent condensation in the tubings and offer the maximal efficiency of HFO ventilation.

3) To have a really efficient heated wall incubator allowing a relative humidity > 70 % at temperatures close to 38 - 38.5 °C.
   It should be equipped with heated air screens at the portholes so that you can open them without upsetting the environmental conditions.
   This in order to allow nursing of very low birthweight preterm infants in conditions of inverted temperature gradients.

Recommendation for Future Research

1. To have an incubator with a built-in monitor able to detect the beginning of intracellular energy failures (integration of the measurements of several temperature probes for instance).

2. To have an i.v. mixture of glucose-amino acids, which closely mimics transplacental fluxes in order to prevent glucose intolerance and intracellular energy failure. To have, beside insulin, other glucose transporter's upregulating factors.

3. To have an easy way of monitoring simultaneously peripheral circulation (via skin heat flux and/or skin blood flow) and ambient temperature. This in order to quantify the response of cardiac patients treated with agents like prostaglandins, catecholamines, vasodilatators, to name a few.

References

1. Brück K (1992) Neonatal Thermal Regulation. In : Textbook of Fetal and Neonatal Medicine. Physiology and Pathophysiology. Polin RA, Fox WW, eds.Saunders Philadelphia, London,Toronto, : 464 - 76.

2. Hey E (1994) Thermoregulation. In : Neonatology: Pathophysiology and Management of the Newborn. Avery GB, Fletcher MA and MacDonald MG. eds. Lippincott, Philadelphia : 357-365.

3. Mann T, Eliott R (1957) Neonatal cold injury due to accidental exposure to cold. Lancet 1: 229:

4. Peliowski A, Finer N.N (1992) Birth asphyxia in the term infant. In : Effective Care of the Newborn. Sinclair J.C and Bracken M.B. Editors. Oxford University Press. Oxford, New York, Tokyo 1992, 257-8.

5. Lunkenheimer PP, Frank I, Ising H, Dickhut HH (1972) Intrapulmonaler Gaswechsel unter simulierter Apnoe durch transtrachealen, periodischen Druckwechsel. Anestesist 22: 232-237:

6. The HIFI Study group (1989) High-frequency oscillatory ventilation compared with conventional mechanical ventilation in the treatment of respiratory failure in preterm infants. N Engl J Med 320: 88-93.

7. Bohn D (1989) High frequency oscillation. Br. J. Anaesth. 63: 16-23 S.

8. Tamura M, Miyasaka K (1991) High Frequency ventilation (letter). Lancet 335: 1354-1355.

9. Clark RH, Gerstmann DR, Null DM, de Lemos RA (1992) Prospective Randomized Comparison of High Frequency Oscillatory and Conventional Ventilation in Respiratory Distress Syndrome. Pediatrics 89: 5-12.

10. Chatelain A, Univ. University Institute for Applied Physics. EPFL Engineering School Lausanne Switzerland. Personal communication May 1994.

11. Sulyok E., Jéquier E., Prod'hom LS (1973) Respiratory contribution to the thermal balance of the newborn under various ambient conditions. Pediatrics 51: 641-650.

12. Micheli JL, Schisler K, Schutz Y, Pfister R, Calame A, Jéquier E (1991) Neonatal adaptation of energy and protein metabolism. Journal of Perinatal Medicine 19: 87-106 suppl.

13. Micheli JL (1981) Métabolisme énergétique et thermorégulation du nouveau-né. Revue de Pédiatrie 18: 341 - 354.

14. Gudinchet F, Schutz Y, Micheli JL, Stettler E. Jéquier E (1982) Metabolic cost of growth in very low-birth-weight infants. Pediat. Res. l6 : l025-1030.

15. Serwer G.A (1992) Postnatal cirulatory adjustments. In : Textbook of Fetal and Neonatal Medicine. Physiology and Pathophysiology. Polin RA , Fox WW, eds. Saunders Philadelphia, London,Toronto, 710 - 20.

16. Hammarlund K, Nilsson GE, Öberg PA, Sedin G (1977) Transepidermal water loss in newborn infants. Acto Paediatr Scand 66: 553-562.

17. Maurer A, Micheli JL, Schutz Y, Freymond, D, Jéquier E (1984) Transepidermal water loss and resting energy expenditure in preterm infants. Helv Paediat Acta 39: 405-418.

18. Sedin G, Hammarlund K, Nilsson G.E, Strömberg B, Öberg PA (1985) Transepidermal water loss in newborn infants. Clin Perinat 12: 79-99.

19. Costarino, AT, Baumgart S (1988) Controversies in fluid and electroly therapy for the premature infant. Clin Perinatology 15: 863-878.

20. Kjartansson, S (1991) Insensible water loss in newborn infants during exposure to non-ionizing radation and cold air. Thesis Uppsala University.

21. Sedin G (1994) Neonatal heat transfer, routes of heat loss and heat gain. Workshop on temperature control, temperature monitoring and optimal thermal environment for sick and low birthweight neonates. Reinhartshausen, October 26-29.

22. Silverman WA, Foreword. In: Sinclair JC, Bracken MB eds. (1992) Effective care of the newborn infant. Oxford: University Press : 5-6.

23. Silverman WA, Fertig JW, Berger AP (1958) The influence of thermal environment upon the survival of newly born premature infants. Pediatrics 22: 876-881.

24. Reynolds EOR (1987) Perinatal brain injury. In: Diseases of the nervous system, 2nd edition. Ashbury AK editor. Blackwell Scientific Publications, Oxford : 12-24.

25. Hope PL, Moorcraft BM (1991) Magnetic Resonance Spectroscopy. Clinics in Perinatology 18: 535-548.

26. Sunehag A, Ewald U, Larsson A, Gustafsson J (1993) Glucose production rate in extremely immature neonates studied by the use of deuterated glucose. Pediatr Res 33: 97-100.

27. Thorens B (1993) Facilitated glucose transporters in epithelial cells. Annu. Rev. Physiol. ; 55: 591-608.

28. Dauncey MJ, Rudd BT, White DA, Shakespear RA (1993) Regulation of Insulin-like Growth Factor Binding Proteins in young growing animals by alteration of Energy status. Growth Regulation ; 3: 198-207.

29. Lafeber HN, Sulkers EJ, Chapman TE, Sauer PJJ (1990) Glucose production and oxidation in preterm infants Pediatr Res: 28 : 153-7.

30. Stefano JL, Norman ME, Morales m.C, Goplerud JM, Mishra OP, Delavoria- Papadopoulos M (1993) Decreased erythrocyte Na+ K+ ATPase activity associated with cellular potassium loss in extremely low birth weight infants with nonoliguric hyperkalemia. J Pediatr ; 122: 276-84.

31. Lucas A, Morley R, Cole TJ, et al. (1990) Early diet in preterm babies and developmental status at 18 months. Lancet ; 335 : 1477-81.

32. Heymann MA, Teitel DF, Liebman J (1993) The heart. In : Klaus MH, Fanaroff MB eds. Care of the high risk neonate. Fourth edition. Saunders, Philadelphia, London, Toronto : 345-373.

33. Sauer PJJ, Dane HJ, Visser HKA (1984) New standards for neutral thermal environment of healthy very low birthweight infants in week one of life. Archives of Disease in Childhood ; 59: 18-22.

34. Micheli JL, Schutz Y, Pfister R, Laubscher B, Calame A, Jéquier E (1993) Fluid requirements, water balance, and metabolism in low birthweight infants. In: Salle B and Swyer P eds. Nutrition of the low birthweight infant. Nestlé Nutrition Series , 32: 43-59.

35. Freymond D, Schutz Y, Decombaz J, Micheli JL, Jéquier E (1986) Energy Balance, hysical Acivity, and Thermogenic Effect of Feeding in Premature Infants. Pediatr Res 1986; 20: 638-645.

36. Micheli JL, Schutz Y, Jéquier E (1992) Protein metabolism of the newborn. In : Textbook of Neonatal and Fetal Medicine - Physiology and Pathophysiology, Polin R.A. and Fox W.W. Editors. Saunders, Philadelphia, London, Toronto : 462-71.

37. Rivera A Jr, Bell EF, Bier DM (1993) Effect of intravenous amino acids on protein metabolism of preterm infants during the first three days of life. Pediatr Res ; 33: 106-11.

38. Lagercrantz H, Slotkin A (1986) The "stress" of being born. Scientific American ; 254 : 92-102.

39. Bell EF (1992) Fluid therapy. In: effective care of the newborn infant. Sinclair J.C. and Bracken M.B., eds. Oxford University Press pp 53-71.

40. Winter JSD (1992) Fetal and Neonatal Adrenocortical Physiology . In : Textbook of Neonatal and Fetal Medicine - Physiology and Pathophysiology, Polin R.A. and Fox W.W. Editors. Saunders, Philadelphia, London, Toronto : 462-71.

41. Stopfkuchen H, Racké K, Schwörer H, et al. (1991) Effect of Dopamine infusion on plasma catecholamines in preterm and term infants. Eur J Pediatr ; 150: 503-06.

42. Weindling AM (1989) Blood pressure monitoring in the newborn. Arch Dis Child ; 64: 444-447.

43. Gill AB, Weindling AM (1993) Randomised controlled trial of plasma protein fraction versus dopamine in hypotensive very low birthweight infants. Arch Dis Child ; 69: 284-287.

Temperature Monitoring

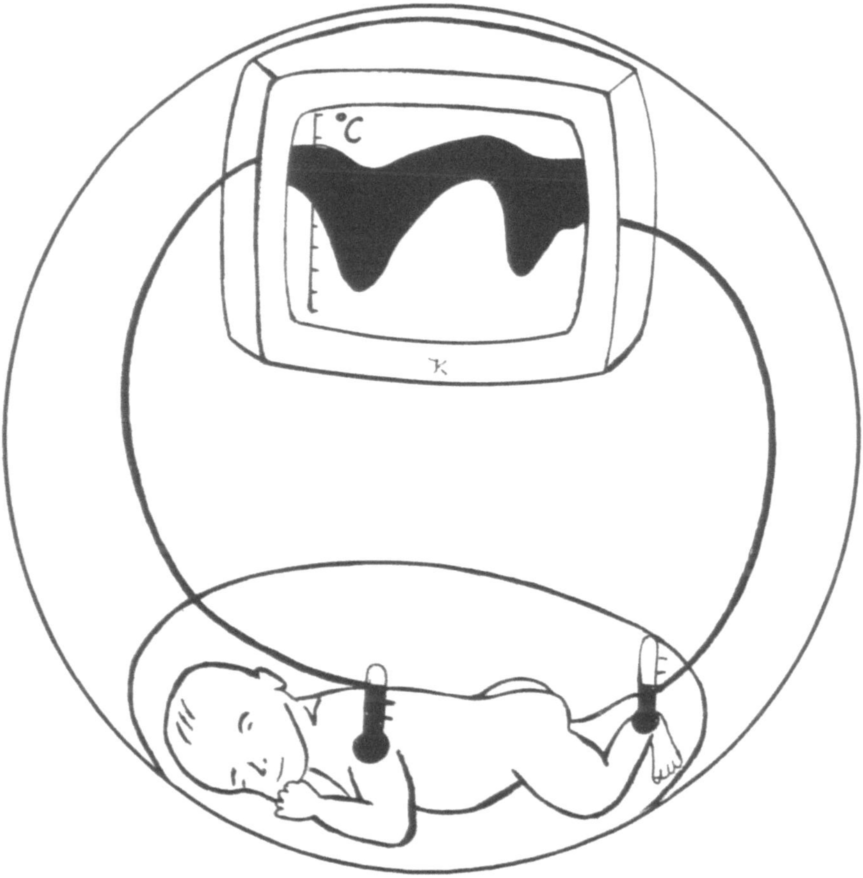

Chapter 5

Temperature Measurements and Distribution of Temperatures throughout the Body in Neonates

GEORG SIMBRUNER

Introduction

Body temperature is one of the most traditionally and generally used terms in medicine and biology but without a precise working definition or thorough physiologic understanding (See below: Definitions). The reasons why body temperatures are measured are rarely considered and consequently it is difficult to choose from the many temperatures and the methods used to measure them (See below: Aims). These methods of temperature measurement have been described in detail and their accuracy and precision characterized but special problems arising in the newborn e.g in measuring esophageal or deep head temperature, have not been addressed (See below: Methods). There is no clear synopsis of temperatures synchronously measured in newborns and in particular in the sick infant who represents a different challenge in terms of measurement of tissue temperature (See below: Profiles in health and disease). Finally, there are no recommendations about which temperature or temperature set, from the many available, should be measured in which patient (See below: Recommendations). This article attempts to answer some of these questions.

Working Definitions of "Body" Temperatures.

There is an infinite number of temperatures which could be measured within body cavities, tissues and on the surface, as there is an infinite number of spatial points in the body's Cartesian grid. Consequently there is a priori no such a thing as "one body temperature", but an a posteriori (after having measured many, many temperatures) temperature which is identical to the integral temperature of the whole body or a segment at a certain time. This temperature we have called the "ideal integral temperature" (of the body or body segment)

Next to the infinite, and thus unmeasurable number of body temperatures, the best representations are temperature profiles; that is series of temperatures along axial or radial distances within tissues which can be represented by mathematical models. Establishing a temperature profile requires a number of measurements within the tissue and one at the surface or maximal radius, i.e. a skin temperature, of that body segment. As an approximation, three temperature measurements (the core or highest temperature, the surface temperature and a temperature in between) may suffice, if a mathematical function, be it monoexponential or polynomial, of the temperature profile is assumed. As a very rough approximation of a profile, a core and skin temperature may be sufficient if the parameters of the mathematical equation are assumed to be known. From a temperature profile the integral (a sort of mean) temperature can be derived. This temperature is obtainable in reality and we have named it the "real integral temperature" or more usually the "integral temperature" (of body or body segment).

The best representation, other than a temperature profile and its integral temperature, is a temperature within the tissue which

a) numerically comes closest to and more or less represents the integral temperature of that body segment at a given time, and
b) changes in parallel to the integral or reference temperature. This temperature we have named "mean temperature" and this can be used for heat balance equations.

Another option is a temperature with its value somewhere between core and skin temperature which, though in absolute terms different from the integral temperature, changes by approximately the same amount as that reference temperature. This we have named the "reference temperature" and it can be used as a reference for the temperature or energy level of the tissue.

Methods for Measuring Temperatures in the Newborn Infant

Various methods have been used to measure temperatures in tissue, their cavities and surfaces. In principle these methods can be grouped into direct and indirect. The former are those where a sensor measures the temperature under zero heat flux conditions which are provided naturally by the body, e.g. rectal or esophageal measurements. There is as yet no special thermosensitive probe available for measuring esophageal temperature in small neonates and we therefore describe one developed using presently available technology.

Temperature Measurement in Particular in the Esophagus

A thermistor-catheter (YSI, 500 Series, Type 520) is inserted into a thin-walled feeding tube with a lumen similar to the outer diameter of the thermistor (e.g. Rüsch, size Nr. 1, Cat.Nr. 334000). The insertion should be without any force to prevent kinking and damage to the expensive thermistor-catheter. The tip of the thermistor-catheter is advanced to a point about 1 cm distal to the end of the feeding tube, so that it sees air but not water or mucus. This sensor can be calibrated using a thermistor of the same mass attached to the outside wall of the feeding tube, to compensate for the thermal resistance offered by the wall of the tube. The error which can arise from feeding tubes with thicker walls is considerable (Table 5.-1)

Tab. 5.-1 : Differences of temperatures inside and outside various feeding tubes

	T difference (° C)
Type A (Rüsch, Nr.1)	0.184 + 0.012
Type B (Rüsch, Nr.2)	0.874 + 0.006
Type C (Sterimed)	0.968 + 0.01

The feeding tube with thermistor-catheter is inserted to a point close to the heart and an approximate distance measured from the alae nasae to the ear and then to the midsternum. This position can be refined by moving the thermistor catheter alone, or the feeding tube and thermistor-catheter together, within the esophagus in steps of 1 cm towards the head or stomach to find the highest temperature, which is assumed to be functionally closest to the heart and therefore to blood temperature [1].

Zero Heat flux Temperature Measurements at the Abdomen and Head

Indirect temperature measurements are those where a sensor measures a temperature at the surface of the body or body segment and where zero flux conditions are artificially produced by either insulation of the site (passive) or by actively heating the site until there is zero heat gradient (active). These methods are called "zero gradient" or "zero heat flux" methods. We prefer the latter method because it is more specific as zero heat flux is the result of a zero temperature gradient. Examples are the zero heat flux method applied experimentally to the abdomen of the newborn piglet [2] or clinically to the trunk [3] and head [4, 5].

The zero heat flux method is based on the fact that a surface temperature approaches a deep tissue temperature if the heat flux at that site is, or approaches, zero. This method is used for measuring deep tissue temperatures non-invasively from the surface. It relies on the artificial creation of zero heat flux conditions in a defined tissue compartment. The passive mode i.e. insulating a particular tissue, creates two compartments (two different heat capacities, two different heat resistances), namely the insulating material and the underlying tissue. If the insulating material does not have too large a heat capacity and is warmed up, its heat storage capacity is considered to be zero and negligible. In the active mode a heating mechanism is regulated so that zero heat flux is achieved and only the tissue compartment has to be warmed (one heat capacity and resistance). In order to obtain a deep tissue temperature the heat flux has to decrease considerably (to at least 10 % of initial value) and better to a negligible amount. If the heat flux is not exactly zero mathematical equations can be used to calculate temperature at zero heat flux. Two correction equations can be used - where the geometry is plane and no heat is generated (Appendix A) and where the geometry is a sphere and internal heat is generated (Appendix B).

Infrared Tympanic Temperature Measurements

Measurement of tympanic temperature has been attempted with various devices but without much success in the newborn. The problem lies in properly focusing the infrared sensor on the tympanic membrane and also in the calibration procedures [6]. Direct tympanic temperature measurement seems more reliably although large differences have been found in two studies in healthy newborns [6, 7]. There is also a danger of perforation of the ear drum and this prevents most investigators from using this method.

Aims to Measure Temperatures Throughout the Body

Temperature is a measure of the energy level of a tissue. Together with the heat capacity and mass, temperature measurement allows the assessment of the amount of heat energy stored in that tissue. The physiologic aims of measuring tissue temperatures are therefore a) to assess heat storage, particularly as a change over time (rate) and b) to monitor the energy level of tissue.

For assessment of heat storage in the body, temperature profiles, or mean temperature of as many body segments as possible, have to be measured accurately. This is important because mean temperature and its rate of change varies from segment to segment, e.g. from trunk to head, particularly in disease states. Also, small changes in mean temperature usually indicate a large amount of change in heat stored. For example, a temperature drop of 0.1 °C per kg tissue mass within 10 min is equivalent to a rate of heat loss of 10 cal/min.kg, which in turn is equivalent to 2 ml/min.kg increase in oxygen consumption. Assuming an oxygen consumption rate of 6 ml/kg.min, the 0.1 °C decrease in temperature, i.e an increased heat loss of 10 cal/kg.min, can only be compensated for by a 33 % increase in heat production, i.e oxygen consumption. This theoretical calculation agrees with empirical data from Ryser and Jequier [8] on the relationship of heat loss to storage in mature newborns. From this viewpoint, any mean body temperature, such as the rectal temperature, would have to be monitored with an accuracy better than 0.1 °C and therefore would be an unsuitable control variable to regulate thermal devices. Consequently measuring more than one core temperatures in a core-shell model is required for assessing heat storage. Very few data are available to indicate how many temperature measurements are required to yield a reasonable assessment of heat storage in the neonate. One core and two shell temperatures as input into a thermal model yielded changes in heat storage which determined the calculated heat production and corresponded surprisingly well with changes in oxygen consumption measured in newborns (1, 9, 10). Most probably the mean temperature of head, trunk and extremities has to be measured for clinically useful assessment of heat storage.

For monitoring the energy level of the tissue, the pathophysiological impact of the tissue temperature on outcome during ischemic-hypoxic events has to be taken into account. The most important tissue to be monitored from this point of view is the brain. Temperature of the brain plays an important role in the regulation of body temperature as well as affecting metabolism, hemodynamics and electrical activity. Over the past years it has become evident that brain temperature plays a crucial role in determining survival and/or brain damage following ischemia and other forms of cerebral hypoxia. A large body of evidence exists that brain temperature determines neurotransmitter response and neuronal damage caused by the hypoxic insult [11, 12]. Brain temperature decreases of 2-3 °C have a significant protective effect on the neuronal damage caused by ischemia. Recently it has been shown that head core, i.e. brain temperature, of hemodynamically compromized newborns can increase by about 1.5 °C [5]. Thus, for monitoring and manipulating brain temperature, a reference temperature measurement (definition see above) to an accuracy of 0.1 to 0.2 °C seems sufficient.

Temperature Profiles in Healthy and Diseased Newborns.

Body temperatures have been measured at numerous sites and in various ways. In human adults temperatures have been measured in body cavities such as the rectum, axilla, mouth, esophagus, trachea [13], vagina (14), meatus acousticus and tympanic membrane [15]; in blood vessels like the aorta and various veins [16]; and in tissues such as bone marrow, liver and brain (17).

The most integral temperature which could be measured in the newborn would be central blood temperature but this has only been reported in one publication [15]. Temperatures near the heart, i.e. in the esophagus, are considered to be closest to a mean body temperature. In the newborn infant temperatures have been measured mainly in the compartments of lower and upper trunk, and rarely in head and extremities. However, most of the investigations only examined two or three temperatures synchronously and the data comes mainly from healthy newborn infants. It has been assumed that healthy neonates were all in a similar clinical state. As shown in Table 5.-2, in healthy newborns the highest temperatures have been measured as head core temperature at the side, where the head is insulated by the mattress upon which it is resting. The next highest are the tympanic, nasopharyngeal and esophageal, followed by rectal and extremity core temperatures. Table 5.-2 also shows that sick newborns have a distinctly different temperature profile. Head core or brain temperature can be lower than esophageal [5, 15] if the brain is damaged either by hydrocephalus or ischemic-hypoxic insults, a condition assumed to be associated with lower metabolism and increased cerebral perfusion.

Head core or brain temperature at the insulated side of the head has been found to be significantly higher (up to 1.5 °C) in newborns with severe shock and decreased cerebral perfusion [5]. Thus low perfusion of the head generally results in high brain, somewhat lower esophageal and even lower rectal temperatures. The esophageal - rectal temperature difference in newborns with congenital heart disease is increased depending on the impairment of hemodynamic performance, i.e cardiac output. It may rise from 0.2 °C in the healthy to 1.2 °C in a newborn with hypoplastic left heart syndrome.

Tab. 5.-2 : Numbers in Parenthesis ( ) indicate the source of data as listed in literature; * indicate own, unpublished data

	Term Newborn Infant			
Head Temperatures	healthy		abnorm hemodynamics	
Intraventricular	> 37.5	?	> 39	?
Brain as Head Core	37.5	[5]	37.8 - 39	[5]
Tympanic	36.8 ; 37.5	[6, 7]	?	
Aural	37.2	[15]	?	
Nasopharyngeal	36.5-37	*	> 37	*
Trunk Temperatures				
Tracheal	?		?	
Esophageal	36.8 ; 37	[5, 8]	37.0-37.5	[5]
Blood	?		?	
Axilla	36.9	[3]	?	
Skin-Mattress (Back, Abdomen)	36.8	[3]	?	
Deep Rectal (5 cm)	37.1	[3]	36.7	*
Rectal	36.9	[3]	36.4	*
Peripheral Core Temperatures				
Calf to .Calf	Peripheral		Tskin + 0.5 °C [1]	

Recommendations for Measuring Temperatures Throughout the Body

The recommendations are dependent on

a) the size of the infant,
b) the health-disease status,
c) the aims of the measurements and
d) technical availability and feasibility.

Table 5.-3 offers recommendations for newborn infants who require artificial heat energy support in the form of an incubator, radiant heater or heated mattress, and these suggestions are grouped according to the number of measuring sites.

Choosing one temperature measurement on the trunk, the esophageal temperature with the method described above is clearly reserved for sick newborns where continuous monitoring may be helpful for clinical management (hemodynamic changes, fever etc). Once feeding tubes with mounted thermistor probes are commercially available and insertion does not affect the infant's care, esophageal temperatures should be monitored. In all newborns, particularly prematures, a continuously measured core temperature of the trunk or abdomen, obtained from a zero heat flux measurement, is sufficient for surveillance of thermal status. Axillary temperature measurement is the simplest form of this core temperature monitoring.

Choosing two temperatures, the combination of trunk core and peripheral skin, allows interpretation of changes in core temperature and introduces the core-peripheral difference as a monitor variable (see contributions of Lemburg, McIntosh and Messaritakis). In future, with more experience on head core temperature, the combination of head core and peripheral skin temperature might prove superior to the trunk core-peripheral skin combination. There is no rational for combining two core temperatures of head and trunk.

Choosing three temperature measurements creates the possibility of gaining important information on the thermal and pathophysiological condition of the infant. This set of temperatures, together with environmental temperature, can be used as the input into a thermodynamic model and yields estimates of metabolic heat loss, production and storage as well as the distribution of circulation [1,9,10]. These variables are required for diagnostic and therapeutic decisions. In addition, such temperature measurements, in combination with a model, would include geometric and environmental factors, thus avoiding the difficulties encountered when one wants to standardize and/or interpret core-peripheral temperature differences. The set of three temperatures, including the head and esophagus as core temperatures, seems to be the best method of monitoring pathophysiological events such as hemodynamic shock and their detrimental effects on the brain.

Tab. 5.-3 : Implications

Implications for Practice
For Newborns, requiring Artifical Energy Support (e.g. Incubator etc) One Temperature Measurement :* Near the heart ! First Choice Tesophageal, if technically feasable Second Choice Tcore /zero Heat Flux, as Tskin - Insulator Pad - Matrass Third Choice Taxillary Two Temperature Measurements : * Near the Heart or Head + Periphery ! First Choice Best Option of One Temperature Measurement + Peripheral Skin Temperature Second Choice Tcore /zero Heat Flux of the Head + Peripheral Skin Temperature Last Choice Best Option of One Temperature Measurement + Tcore /zero Heat Flux of the Head
Recommendations for Future Research
Three Temperature Measurements : * Near Heart, Head & Periphery ! First Choice for Energy Balance Equation Best Option of One Temperature Measurement + Central & Peripheral Skin Temperature First Choice for Tissue Temperature Tcore /zero Heat Flux of the Head + Tesophageal + Peripheral Skin Temperature

References:

1. Simbruner G. Thermodynamic models for diagnostic purposes in the newborn and fetus. Facultas Verlag, Vienna, ISBN 3-85076-133-9

2. Dollberg S, Xi Y, Donnelly MM (1993). A noninvasive transcutaneous alternative to rectal thermometry for continuous measurement of core temperature in the piglet. Pediatr Res; 34: 512-517

3. Mayfield SR, Bhatia J, Nakamura KT, Rios GR, Bell EF (1994). Temperature measurements in term and preterm neonates. J Pediatr; 104: 271- 275

4. Templeman MC, Bell EF (1986). Head insulation for premature infants in servocontrolled incubators and radiant warmers. AJDC ; 140: 940-942

5. Simbruner G, Nanz S, Fleischhacker E, Derganc M (1994). Brain temperature discriminates between neonates with damaged, hypoperfused and normal brains. Am J Perinatol ; 11: 137-143

6. Johnson KJ, Bhatia P, Bell EF (1991). Infrared thermometry of newborn infants Pediatrics ; 87: 34-38

7. Mayfield SR, Nakamura KT, Bhatia J, Rios GR, Bell EF (1984). Tympanic membrane temperature in term and preterm neonates. Early Hum Dev: 9 : 241-247

8. Ryser G, Jequier E (1972). Study by direct calorimetry of thermal balance on the first day of life. Europ J Clin Invest; 2: 176-187

9. Simbruner G, Timischl W, Karigl G, Zeitlhuber U (1981). Ein thermodynamisches Modell zur Beurteilung des Sauerstoffverbrauches und des peripheren Blutflusses beim Neugeborenen. Anästhesist ; 30: 358 - 363

10. Simbruner G, Karigl G, Hirn B, Popow Chr, Timischl W (1982). Die Bestimmung des Sauerstoffverbrauches bei Neugeborenen aus Temperaturmessungen unter Zuhilfenahme eines thermodynamischen Modells. Wiener Klin. Wochenschr.; 94(11) : 280-285

11. Busto R , Dietrich WD, Globus MY (1987). Small differences in intra-brain temperature critically determine the extent of ischemic neuronal injury. J Cerebr Blood Flow Metab; 7: 729-38

12. Sterz F, Safar P, Tisherman S, Radovsky A, Kuboyama K, Oku K (1991) Mild hypothermic cardiopulmonary resuscitation improves outcome after prolonged cardiac arrest in dogs. Crit Care Med ; 19: 379-389

13. Yamakage M, Kawana S, Watanabe H, Namiki A (1993) The utility of tracheal temperature monitoring. Anesth Analg ; 76: 795-9

14. Abrams RM, Royston JP, Humphrey SJE, Wolff HS (1980) Thermal features of the female axilla J Appl Physiol.: Respir, Environm, Exercise Physiol. ; 49(6): 1042-1046

15. Stratton T (1977) Aural temperature. Arch Dis Child ; 52: 865-869

16. Eichna LW, Berger AR, Rader B.Becker WH (1951) Comparison of intracardiac and intravascular temperatures with rectal temperatures in man. J Clin Invest ; 30: 353-359

17. Mellegard P, Nordström CH (1990) Epidural temperature and possible intracerebral temperature gradients in man. Brit J Neurosurg ; 4: 31-38

Chapter 6

Thermal Monitoring of very Preterm Infants. Which Temperature should be Measured ?

PETER LEMBURG

During intensive care in VLBW there are very often problems of insufficient weight gain and growth in spite of sufficient support with carbohydrates, fat, and protein [1-6]. At the same time we could observe that peripheral vasoconstriction was present with incubator care at "neutral-temperature-conditions". The babies showed signs of thermal stress. We monitored rectal temperature and felt that this was insufficient for controlling the babies optimal thermal conditions. The questions is, whether intensive care at neutral-temperature-conditions is appropriate for very sick VLBW-babies ?

Until now there have been no continuous temperature recordings at different sites of the body in VLBW-babies during intensive care at postnatal day one, three and seven and including recording of ambient conditions and intensive care procedures. We used modern computerized data acquisition devices and software to handle this huge amount of data.

Our study was not performed with the aim of reaching comfort conditions by altering the ambient thermal conditions, but we were able to find situations during care in which the babies were nursed under conditions with a delta-T between 1 and 2 °C. Most of the time the babies were not within "thermal-comfort"-conditions but in "neutral-temperature-range" [7].

Method

Temperature measurements from interscapular skin-mattress (SM), processus mastoideus, mid-forehead, abdomen above liver and the back of the forefoot (FF) were made on 14 ventilated VLBW-babies during intensive care. On day 1, 3, and 7 body- and ambient incubator-air-temperature (IAT) were monitored continuously along with ambient incubator-air-humidity (IAH). Inspiration-air-temperature of the ventilator, temperature of the incubator walls and the room air were continuously collected. Rectal temperature and heart-rate were recorded simultaneously. Changes of babies behaviour were manually noted.

Results

Incubator care using standard neutral temperature settings was associated with a distinct temperature gradient between central (core = SM) and peripheral (FF) surface-temperature. The gradient "delta T" was up to 6 °C in maximum (Fig. 6.-1).

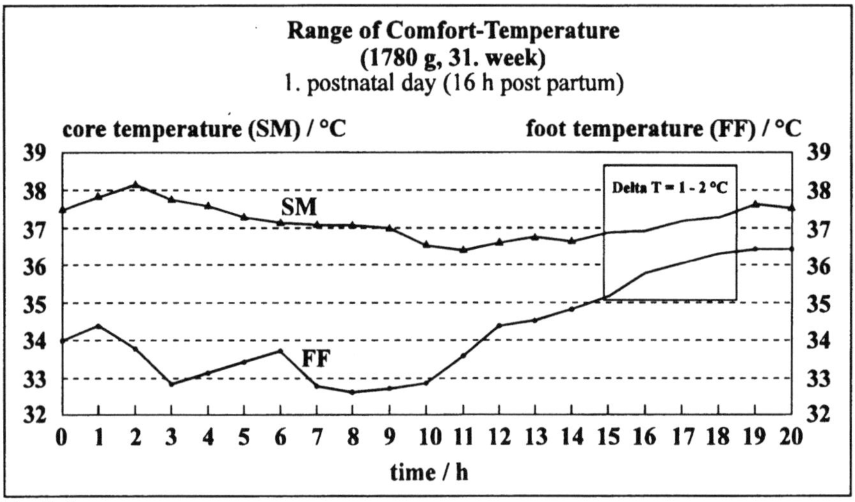

Fig. 6.-1

This was followed by increased vasomotor tone which may be undesirable during intensive care, especially on day one, a few hours after birth when dealing with severe problems of respiratory distress. To overcome this the incubator air-temperature was raised by the nursing staff until vasoconstriction and signs of thermal stress disappeared. Core-temperature (SM) was then about 1 degree higher than usually published, ie 37,5 to 38,5 °C, with the "delta T" showing a gradient of 1,5 ± 0,5 °C (Fig. 6.-2).

Comparison of the 6 body-temperature-recordings mentioned above revealed that monitoring rectal temperature during intensive care was of least value. One problem is to define the precise localization of measurement within the rectum. There are big differences, up to 3 °C, depending upon the distance of the thermometer tip from the anus. It is an invasive procedure and may perforate the rectum [8-11]. It is influenced by the temperature of the blood returning from the limbs. Rectal temperature monitoring revealed increased gradients to core temperature if the sensor was not correctly positioned. If recorded a distance of 25 mm from the anus there was a gradient to core temperature of up to 1 °C. The probe may be dislodged during passage of stools. Severe peripheral vasoconstriction from any cause (shock etc.) resulted in an increased gradient to core-temperature, up to 1 °C maximum.

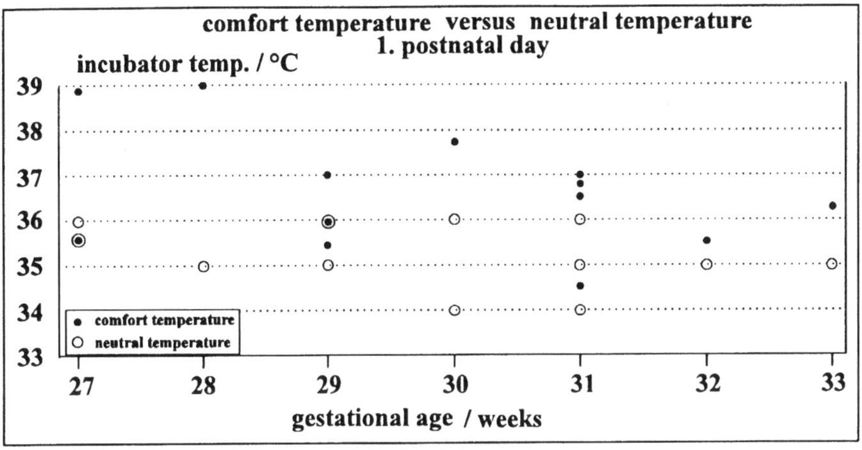

Fig. 6.-2 : „Comfort-temperature" compared to neutral temperature (derived from birthweight and gestational age, Obladen 1989) for incubator-air temperature setting. „Comfort-temperature" as recorded from 14 cases of VLBW is 1.4 °C (mean) higher than calculated neutral temperature.

The often and rapidly changing clinical condition of the VLBW babies demands two-point temperature-monitoring, both central and peripheral. The temperature gradient, delta T, correlates with the preset IAT, IAH, and the degree of vasomotor tone of the baby.

We observed a reciprocal relation between IAT, IAH, and delta T. High IAH (>75rel. %) and decreased IAT or low IAH (< 60rel. %), and elevated IAT were followed by the same delta T. Within the range of delta T = 1-2 °C, there was no evidence of increased vasomotor tone and this correlation was observed during different situations in intensive care.

The range of delta T (1-2 °C) was achieved with a mean increase of IAT of approximately 1,5 °C above the usual standard neutral incubator-air-temperature-settings (Fig. 6.-3). There are however no absolute values for incubator air temperature which can be recommended for all babies and these will depend on the clinical condition in individual cases.

We found SM - temperature most useful for monitoring core temperature [12]. It is non invasive and there was no gradient to esophageal temperature. If disposable nasogastric tubes with incorporated temperature-sensor become available this may be an alternative to SM.

SM-temperature showed a gradient of up to 1 °C compared to forehead-temperature, using a well insulated sensor, and 0,5 °C to the processus mastoideus which was dependent on the position of the head and difficult to insulate. Abdominal skin-temperature was up to 0,5 °C lower than SM-temperature. It proved not very useful if a radiant heater or phototherapy were in use.

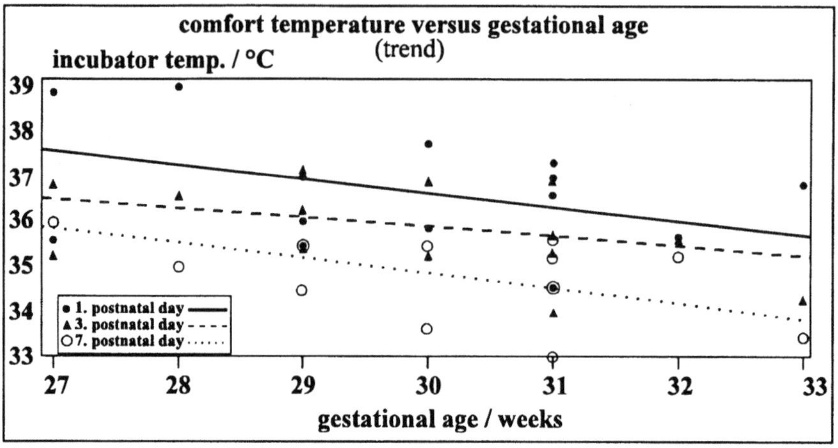

Fig. 6.-3 : "Comfort-temperature" as recorded from 14 cases of VLBW at postnatal day 1, 3, and 7: The incubator-air-temperature for achieving comfort-temperature-conditions including air-humidity of 70 % decreases from day 1 to day 7.

VLBW babies showing delta T within the optimal range of 1-2°C, and without peripheral vasoconstriction, had a core temperature (SM-temperature) of 37,5 up to 38,5 °C (mean). No increase in heart rate above normal was observed. The IAT and IAH necessary to reach this condition of the baby we have called the "comfort-temperature and -climate".

Conclusion

Throughout the course of the study we had the impression that during intensive care under conditions of thermal comfort with continuous two temperature monitoring less problems of care occurred compared to care under neutral temperature conditions. We don't know the influence on morbidity and mortality rates but this is not known for neutral temperature nursing either. Further research is needed.

It was interesting to see how this observational study changed to an interventional one. The nursing and medical staff became more aware of the problem of thermocontrol in VLBW-babies and altered their practice to minimise problems. This is now an objective of an ongoing study.

Implication for Practice

1. Perform two-point-temperature-measurement, central and peripheral

2. Attach central temperature sensor:
   -skin mattress (non invasive, interscapular)
   -esophageal (invasive, combined with nasogastric tube)

3. Attach peripheral temperature sensor to the sole of the foot (non isolated sensor)

4. Try to achieve a delta T according to gestational age and birth-weight and postnatal age.

5. Incubator air humidity should be >75 rel % (minimize evaporative water and heat loss)

6. Avoid rectal temperature measurement (invasive, not reliable, not representative for thermal condition of the VLBW)

7. Avoid increases in peripheral vasoconstriction

8. Monitor behaviour of the baby (restlessness may indicate thermal discomfort)

9. Monitor inspiration gas temperature. Keep near central body temperature and > 80 rel % humidity

References:

1. Day R, Caliguiri L, Kamenski C, Ehrlich F (1964) Body temperature and survival of premature infants. Pediatrics 34: 171-181.

2. Glass L, Silverman WA, Sinclair JC (1969) Relationship of Thermal Environment and Caloric Intake to Growth and Resting Metabolism in the Late Neonatal Period. Biol Neonat 14: 324-340.

3. Glass L, Silverman WA, Sinclair JC (1971) Food, temperature, and head growth in neonates. Lancet 1: 1186-1187.

4. Glass L, Lala RV, Jaiswal V, Nigam SK (1975) Effect of thermal environment and caloric intake on head growth of low birthweight infants during late neonatal period. Arch Dis Child 50: 571-573.

5. Glass L, Silverman WA, Sinclair JC (1968) Effect of the thermal environment on cold resistance and growth of small infants after the first week of life. Pediatrics 41(6): 1033-1046.

6. Sauer PJ, Dane HJ, Visser HKA (1984) Longitudinal Studies on Metabolic Rate, Heat Loss, and Energy Cost of Growth in Low Birth Weight Infants. Pediatric Res 18(3): 254-259.

7. Opladen M (1989) Neugeborenenintensivpflege, Grundlagen und Richtlinien. 4. Auflage, 35-43.

8. Fonkalsrud EW, Clatworthy H. William J (1965) Accidental perforation of the colon and rectum in newborn infants. New England Journal of Medicine 272(21): 1097-1100.

9. Greenbaum EI, Carson M, Kincannon WN, O'Loughlin BJ (1969) Rectal thermometer-induced pneumoperitoneum in the newborn Pediatrics 44(4): 539-542.

10. Merenstein GB (1970) Rectal perforation by thermometer. Lancet 1: 1007.

11. Wolfson JJ (1966) Rectal Perforation in Infant by thermometer. Am J Dis Child 111: 197-200.

12. Gregory E Lamont (1987) The Thermal Control of Newborn Babies in Developing Countries. Thesis (MS) University of Oxford.

Chapter 7

Experiences with Thermal Monitoring, Influence of Neonatal Care and how should it be Monitored

NEIL MCINTOSH / ANDREW WILMSHURST / JANE HAILEY

For the last 10 years we have had experience of thermal monitoring using a computerised cot monitoring system [1-6]. This monitoring system has been crucial in developing our own understanding of the problems of thermal monitoring and has also given us certain insights about the management of the thermal environment of very low birthweight infants suggesting ways of reducing thermal stress in such babies. This paper will review some of the insights we have gained. We will assume that it is accepted that the improper management of the thermal environment of very low birthweight or extremely low birthweight infants will increase morbidity and possibly mortality as shown many years ago [7-10]. It is likely that any infant who has thermal control mechanisms (i.e. is not functioning as a poikilotherm) will utilise energy to maintain its body temperature if stressed by an environmental temperature outside the neutral thermal range. Such energy is utilised at the expense of growth and the smaller the infant the more critical growth is for the infant's survival.

Monitoring System

In 1986 we published information about a computerised monitoring system that was written in *Basic* on an Apple II microcomputer and was able to collect continuously up to 16 channels of physiological and environmental data sampling from the routine infant monitors [1]. The programme was quite slow and the hardware was situated outside the nursery so that the results did not influence the infants management. We were able though to investigate the thermal control of infants managed in single walled incubators comparing air mode incubator control with servo mode control (vide infra) [2] and were also able to do some work on incubator humidification [3]. We believed that our monitoring system was clinically valuable so we rewrote the software in *C* (much faster) and introduced a network of microcomputers (RM Nimbus), each connected to the monitors of 1 intensive care baby and with a server and doctors station at a remote site - the MARY system (named after the ward clerk that we wanted to manage it) [4]. Within 24 hours of the system's introduction in the SW Thames Regional Intensive Care Unit at St George's Hospital, London, it was evident to the nursing staff that extremely low birthweight infants became thermally unstable when any handling procedures occurred (Fig 7.-1). We attempted to more formally evaluate the effect of routine nursing procedures on preterm infants and recorded our findings in 1991 [5]. Our system was again upgraded - rewritten in *C++* and installed on a network of Dell 386 microcomputers on the 16 intensive care cots of the SE Scotland Regional Intensive Care Unit at the Simpson Memorial Maternity Pavilion in Edinburgh [6].

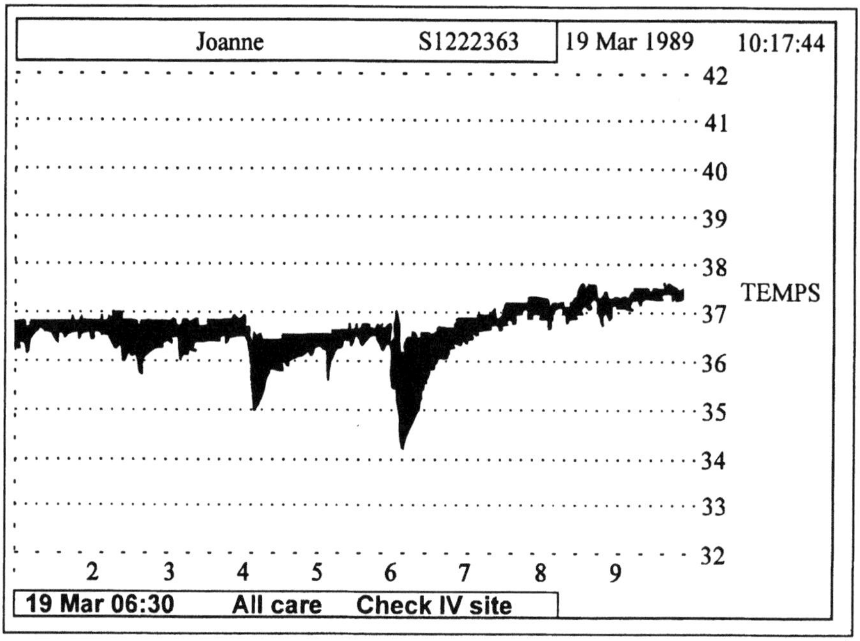

Fig. 7.-1 : Temperature instability in an infant weighing less than 1000 g during a routine care procedure lasting 10 minutes. The infant remained within the incubator during the procedure.

Incubator Temperature Control

In 1985 we studied the effect of the mode of incubator temperature control on the stability of temperatures of very low birthweight infants. 22 consecutive infants of birthweight less than 1500 g were nursed naked in an incubator (the common practice at the time). The infants were covered with a single layer of bubble plastic and the incubator (Vickers 79) was humidified with water in a heated tray in the incubator base. The first 12 infants (mean weight 980 g) were managed by proportional servo control and the subsequent 10 (mean weight 1000 g) with air mode control - the temperature being chosen with reference to the charts of Hey and Katz [11].

Abdominal skin and toe temperatures, as measures of central and peripheral temperatures respectively were measured via Simonsen and Weel 8000 series monitors. Continuous rectal temperatures were not recorded in the unit because of the problems associated with rectal probes in very small neonates. The data was not readily available to the staff, so that decisions on temperature control were not influenced by the study.

Results

The results were derived from more than 700 hours of continuous monitoring in each of the 2 groups of patients. Figs 7.-2 and 7.-3 are typical computer graphs of similar infants in incubators controlled by servo and air mode respectively. These graphs show the great variability of both the ambient and peripheral temperatures in the servo controlled group. Decrease in the abdominal temperature was needed to activate the proportional temperature control of the incubator, by which time the toe temperature had fallen quite markedly. The delay in switching on the heater with a cold abdominal temperature was matched by the delay in switching it off as the abdominal temperature increased, resulting often in an overshoot of the infants temperature above the set value.

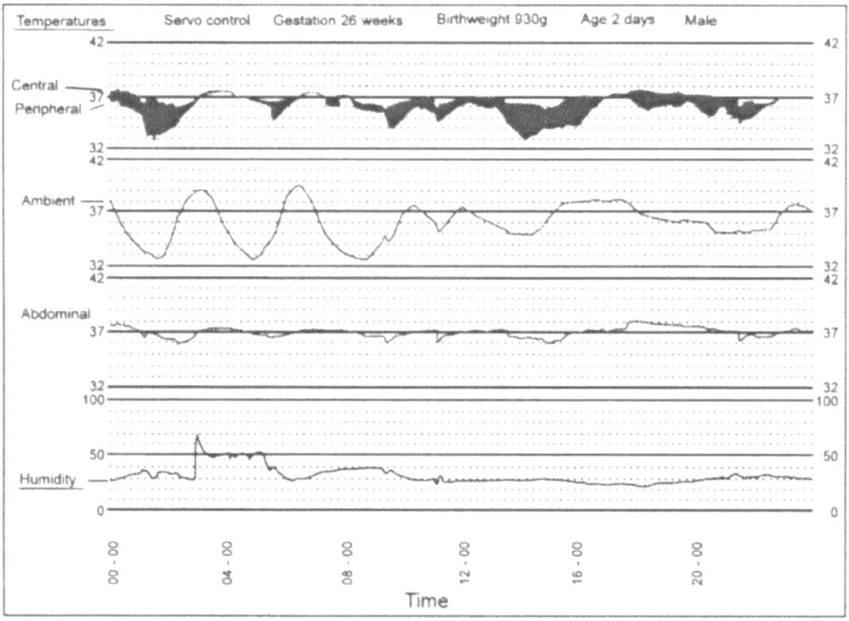

Fig. 7.-2 : Typical 24 hour graph showing fluctuation of central-peripheral temperature with ambient temperature in an infant using servo control. Note the stable abdominal temperature.

For each infant, the variance of the mean daily central-peripheral temperature difference (delta T) was used as a measure of temperature stability. Figure 7.-4 shows for each group the mean (SD) variance of the delta T for each of the first 4 days of life. It can be seen that with servo control there are wide fluctuations in the delta T particularly over the first 2 days of life but that in air mode control the differences are much less suggesting that the infants were much more thermally stable. The change over the first three days of life suggests a quite rapid maturation of temperature control.

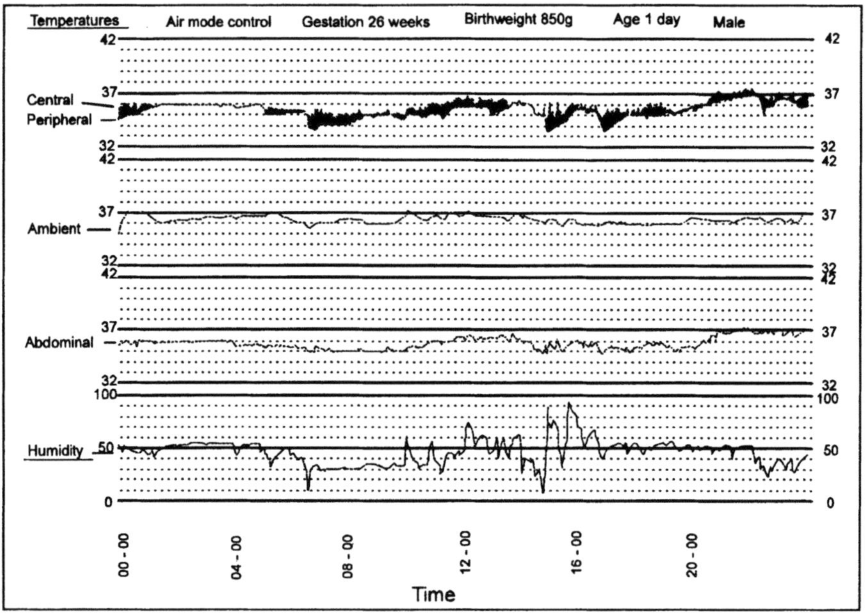

Fig. 7.-3 : Typical 24 hour in an infant using air mode control.

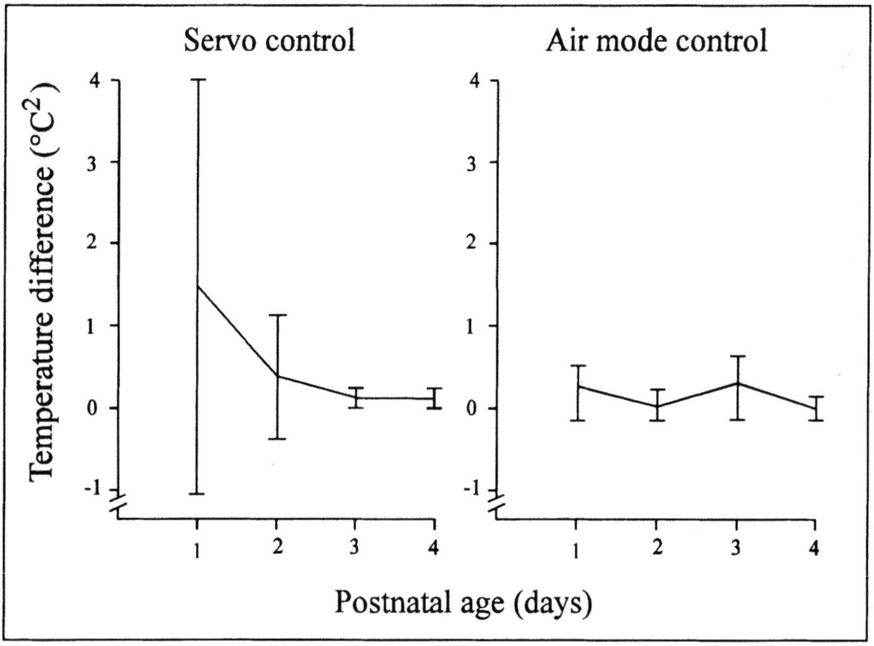

Fig. 7.-4 : Variance (mean (SD)) of central peripheral temperature difference.

Discussion

In this study, computer based thermal monitoring allowed assessment of the *routine management* of infants receiving intensive care. It showed that although servo control was an attractive idea, it did not work when the servo drive was (as suggested by the manufacturers) using an abdominal thermistor. Indeed there was good evidence that this mode of control subjected the small patients to considerable thermal stress, particularly in the first 2 days of life. If servo control were to be more effective it would be necessary to use a more sensitive drive mechanism - possibly a more peripheral temperature e.g. deltoid temperature. Activation would then occur before the baby was subjected to thermal stress. Alternatively the widening of the delta T itself might be used as the servo drive. We have not tested either of these with our monitoring system and in the meantime it would seem appropriate to recommend the use of air mode control for small babies in incubators.

Nursing Care Procedures

With the move of computer monitor from remote research room to the individual babies cotside the nursing staff were able to see the trend of the babies physiology and in particular the effect of care procedures on this physiology. Within 24 hours of its introduction, patterns such as that shown in figure 7.-1 were seen, particularly in the extremely low birthweight group. To assess the effect of these care procedures in more detail we embarked on a review of the thermal effects of nursing care procedures which was published in 1991 [5].

The monitored thermal data from the MARY system [4] was assessed from 25 very low birthweight infants admitted for intensive care. The infants were divided into those above (n=9) and below (n=16) 1100 g. All infants were routinely nursed naked and covered by a single layer of bubble plastic in closed air mode controlled incubators. Temperature probes were attached to the abdominal skin and the sole of the foot to measure core and peripheral temperatures respectively. Total care nursing procedures that were carried out during the first week of life were monitored. These consisted of mouth care, measurement of axillary temperature, nappy change, turning, heel pricking for measurement of blood glucose and rotation of transcutaneous probe sites. If the infant was ventilated, physiotherapy and endotracheal suction would often be included. This routine, performed four to six hourly, took about 20 mins to complete but could vary from 15 mins to 45 mins depending on the nurse and the tasks required.

Results

The babies ranged from 510-1500 (median 920) g with a gestation 24-31 (median 27) weeks. A total of 249 care procedures were analysed from the first week of life. Figure 7.-5 shows the effect of a typical care procedure. The delta T increases significantly with the care procedure (mean increase 0.6 degrees [range 0-2.7]) though both core and peripheral temperatures fell (core temperature by a mean of 0.7 degrees [range 0-1.7] and peripheral temperature by a mean of 1.3 degrees [range 0.2-3.0]). Recovery to the temperature before the procedure was gradual and could take as long as 2 hours. In the infants less than 1100 g there was a strong negative correlation between the fall in peripheral temperature and the birthweight ($r=-0.86$, $p<0.0001$) but for the larger group there was no significant correlation with birthweight and for both groups there was no correlation with gestation.

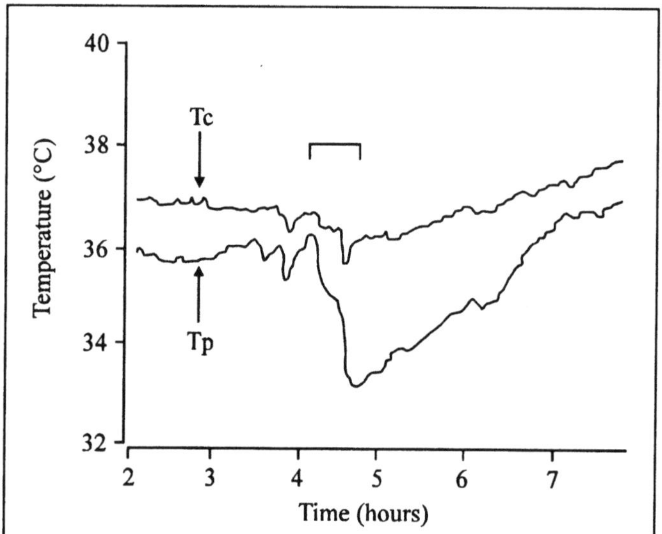

Fig. 7.-5 : Effect of typical care procedure on 680g infant.
T_C = peripheral temperature

Discussion

Despite our attempts to maintain babies within the ideal thermal environment, there was an unavoidable fall in incubator temperature and humidity when portholes were opened for nursing care. This caused a fall in both core and peripheral temperatures (with increasing delta T) particularly in the smallest infants where the surface area to body weight ratio is greatest and the skin is most immature leading to larger evaporative transcutaneous fluid losses which occur even in humidified incubators. The care procedures have now been modified to minimise fluctuations in temperature. For small infants the nurses increase the incubator temperature by one degree before their total care routine, and try to do the various tasks as quickly as possible keeping a close watch on the monitor screen during the process. The visual feed back provided by the computer is important in reducing the thermal instability and possibly the metabolic demands on the baby. It is possible that more recent double shelled incubators, particularly those providing an air curtain will lead to much less thermal stress during such care procedures. We are not aware that this has been specifically tested.

Differential Temperatures as a Measure of Illness

From the above data we can convincingly say that cold stress leads to a fall in both core and peripheral temperature with a widening delta T because of the greater peripheral temperature fall. With heat stress the peripheral temperature will become very close to the core temperature before the core temperature increases and on some occasions, e.g. when the peripheral heat source is very strong, the peripheral temperature will exceed that of the core. This last feature may also be seen when the peripheral probe becomes detached from the baby and lies exposed to the heat source.

We have been aware that a wide delta T may be seen in situations of hypovolaemia - in this case sometimes associated with a fast heart rate and a low blood pressure (both of which can be simultaneously displayed on the computer screen - figure 7.-6). A poor urine production rate and a poor peripheral capillary return and an increasing acidosis may also be seen in this situation, although the latter 2 features may also be seen with a developing bacterial infection.

Tab. 7.-1 : Pattern of physiological response to physiological and environmental stress.

problem	**pT**	**cT**	**Δ T**	**HR**	**BP**	**pH**	**Cap Return**	**WBC (N)**
early cold stress	↓	-	↑	-	-	-	slow	-
late cold stress	↓↓	↓	↑	-	-	acidosis	v slow	-
early heat stress	↑	↑	-	-	-	-	-	-
late heat stress	↑↑	↑↑	O/-	↑	-	-	-	-
early hypovolaemia	↓		↑	↑	-/↓	acidosis	slow	-
late hypovolaemia	↓↓		↑↑	↑↑	↓	acidosis +	v slow	-
infection	↓	↑	↑↑		-/↓	acidosis +	slow	↓/↑

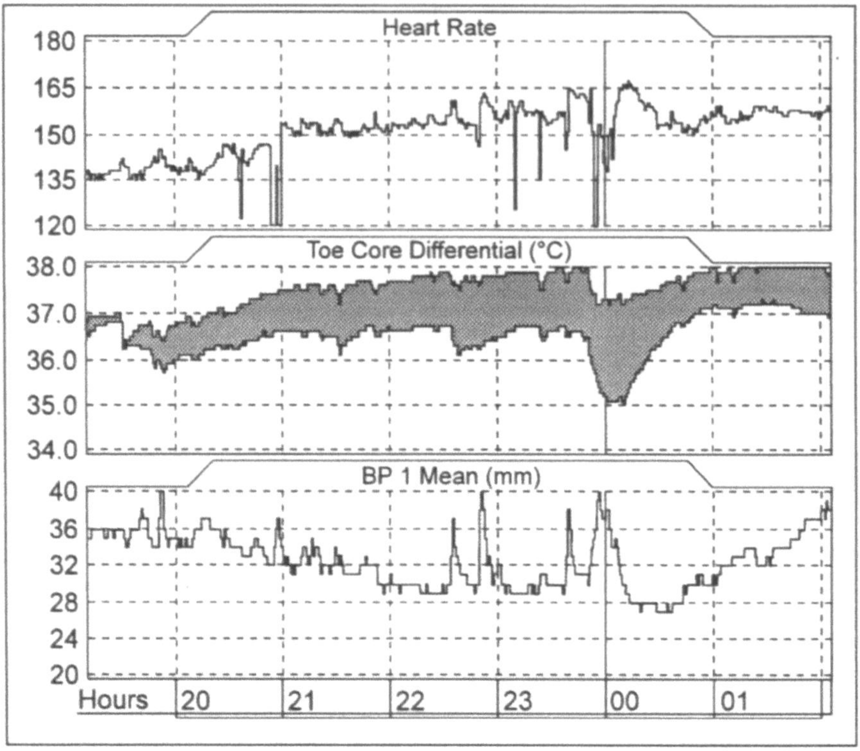

Fig. 7.-6 : Increasing central-peripheral temperature difference with rising heart rate and falling blood pressure, typical of developing hypovolaemia in a preterm baby.

Finally it is well recognised in older patients that infection causes a rise in the central temperature and sometimes a fall in the peripheral temperature. Messaritakis et al have clearly shown that this also happens in full term infants and in a proportion of preterm infants, though not all [12]. Their data will be elaborated in the next paper.

Audit of Differential Temperatures in Extremely Low Birthweight Infants

An audit of temperature control and its management was undertaken by examining the computer records and medical and nursing notes of twenty neonates born in 1994 with a birthweight under 1000g.

Patients

Twenty neonates born in 1994 were audited. A median weight of 750g was chosen and ten babies were audited either side of this weight. The babies were selected on the basis that :

a) Temperature information was available
b) Birthweights were closest to 750g

Tab. 7.-2 : Neonates included in the study

	Group A	Group B
Number (n)	10	10
Weight (range, g)	550-735	755-885
Weight (mean, g)	569	821
Gestation (range, wks)	23-29	24-30
Gestation (mean, wks)	26.4	28.3

The records of the babies were audited for 72 hours unless the baby died before this period had been completed. Three babies in Group A and three in Group B died before three days of monitoring had been completed. The temperature monitoring of one further baby in Group B was stopped after 66 hours as the baby's condition was felt to be stable.

Methods

For each baby the computer records were analysed and the following three abnormal temperature variables were recorded.

1) Core temperature < 36 degrees centigrade
2) Peripheral temperature < 35 degrees centigrade
3) Peripheral-core differential temperature > 2 degrees

An 'Abnormal Temperature Event' (ATE) was defined as an occasion where one, or a combination of more than one of these abnormal variables were recorded. The duration of these events was recorded together with information about possible precipitating events and actions taken. This information was obtained from the MARY computer comments record (comments recorded contemporaneously by Nursing and Medical staff) and by reviewing the nursing and medical written records.

ATE's were divided into three categories:

1) 1st order - one temperature variable abnormal
2) 2nd " - two " " "
3) 3rd " - all " " "

The duration of the event was measured from the time that the first variable became abnormal until all variables had returned to normal. ATE recording was not commenced until the time after admission when all the variables had become normal. Thus temperature stress associated with initial resuscitation and transport from Labour Ward were not audited. In the babies who died before three days were completed, no temperature events were analysed in the three hours prior to their deaths as it was felt that during this period the temperature control might be greatly altered and might skew the data recorded from other periods.

In addition, we recorded the duration of time that each baby spent with a core temperature greater than 37.5 degrees and 38 degrees and also the number and duration of ATE's where the peripheral-core differential was greater than 3 degrees.

Results

Table 7.-3 : Mean no. of ATE's per 24 hours

Group	1st Order	2nd Order	3rd Order*	Total
A	1.2	1.4	0.04	2.7
B	2.1	1.7	0.09	3.8
A+B	1.6	1.6	0.07	3.2

* Only three 3rd order events occurred during the study period - one in Group A and two in Group B - these are excluded from the remaining analysis of the results.

Tab. 7.-4 : Mean Duration of ATE's (Mins) with 10th and 90th centiles

Group	1st Order		2nd Order		p value for difference
A	7.5	(5 - 36.4)	24	(4.6 - 254)	0.0156
B	4	(2 - 21.3)	19	(3.2 - 237)	0.0001
A+B	7	(2 - 32.1)	21	(3.5 - 250)	0.0001

Tab. 7.-5 : Daily Duration of ATE's per 24 hours (Mins, % daily time)

Group	1st Order		2nd Order		3rd Order		Total	
A	20	(1.4 %)	82	(5.7 %)	2	(0.16 %)	104	(7.3 %)
B	18	(1.2 %)	110	(7.4 %)	13	(0.93 %)	138	(9.6 %)
A+B	19	(1.3 %)	95	(6.6 %)	8	(0.54 %)	121	(8.4 %)

The temperature events are analysed separately below. The percentage of events precipitating action and the percentage with an obvious explanation for the event are recorded.

Tab. 7.-6 : First Order Events

Group	Reason	low pT	low cT	incr. deltaT	Action
A	32 %	43 %	46 %	11 %	32 %
B	24 %	39 %	46 %	15 %	17 %
A+B	27 %	41 %	46 %	14 %	24 %

The First Order events were also analysed to show what combination of temperature abnormalities precipitated action. In this group the actions taken consisted of increasing the incubator temperature, giving fluid volume (either plasma protein substitute or Red Cell Concentrate or a combination of both).

Tab. 7.-7 : First Order Events where action taken (n=17)

Abnormal variable	Fluid given	Incr. Inc T	Fluid and incubator
pT<35	2	2	1
cT<36	2	7	-
deltaT>2	-	3	-

Considering the second order events similarly.

Tab. 7.-8 : Second Order Events

Group	Reason	pT+cT	pT+deltaT	Action
A	79 %	73 %	27 %	58 %
B	46 %	41 %	59 %	32 %
A+B	62 %	56 %	44 %	45 %

The Second Order events were also analysed to show what combination of temperature abnormalities precipitated action.

Tab. 7.-9 : Second Order Events where action taken (n=31)

Abnormal variable	Fluid given	Incr. Inc T	Fluid and Incubator
pT+cT	3	11	4
pT+deltaT	2	7	4

There is thus a clear suggestion that abnormality in 2 variables is perceived as being of more significance and more action is precipitated.
There were only 3 third order events, 2 without obvious cause which were both treated with both fluid and increased incubator temperature, and the other was precipitated by long line insertion - no specific action was taken.

If the First and Second Order events are analysed on the basis of whether action was taken or not, the following results are seen.

Tab. 7.-10 : First Order Events where action taken

	Group A		Group B	
	Number	mean duration	Number	mean duration
Action taken	9	32mins	8	5mins
No action	19	9mins	38	9mins
Mann-Whitey of difference		p=0.004		p=0.783

Tab. 7.-11 : Second Order Events where action taken

	Group A		Group B	
	Number	mean duration	Number	mean duration
Action taken	19	69mins	12	112mins
No action	14	41mins	25	42mins
Mann-Whitey of difference		p=0.344		p=0.026

We presume in retrospect that action was taken because of a perception that the thermal stress was more significant. This is certainly suggested from the duration of the problem.

As a further measure of temperature stress, the length of time that each baby's core temperature was greater than 37.5 degrees and 38 degrees was recorded. This is displayed below for each group expressed as a percentage of the length of time of monitoring.

Tab. 7.-12

Group	cT>37.5	cT>38.0
A	15.8 %	3.8 %
B	12.2 %	1.7 %

In addition two babies were recorded as having a delta T greater than 3 degrees. Both these babies were in Group B. There were 3 episodes, the durations were 36, 22 and 11 minutes. All three episodes were associated with the insertion of long lines.

Implication for Practice

1. A central and peripheral temperature should be measured at regular and frequent intervals.

2. The results should be graphed.

3. A low core temperature indicates significant thermal stress which should be treated with a raised environmental temperature.

4. A wide delta T (greater than 2 degrees probably and greater than 3 degrees definitely) is abnormal. A cause should be looked for.
   Possibilities could be
   - Hypovolaemia
   - Environmental cold stress
   - Catecholamine infusion
   - Infection (usually with raised core temperature)
   - Persistant Ductus Arteriosus

Recommendation for Future Research

Questions

1. What is the appropriate core temperature to aim for?
   37.9 ± 0.3°C (ie intrauterine temperature)

2. How do you differentiate the causes of a wide delta T?

3. Is there an optimal rewarming speed?

Collaborators in Work: PJ Badger, CA Bass, DA Ducker, IA Laing, AJ Lyon, Q Mok, R Ross Russell, KA Wheeler. Nursing staff at St Georges Hospital, London and the Simpson Memorial Maternity Pavilion, Edinburgh.

References:

1. Bass CA, Smith JS, Ducker DA.(1984) The use of a computer linked monitoring system for the continuous collection of physiological and environmental data from neonates undergoing intensive care. Chapter 7 pp40-48. Proceedings of the 2nd International Conference on Foetal and Neonatal Physiological Measurement. Oxford, 2-4 ed Rolfe P, Butterworths, London 1996.

2. Ducker DA, Lyon AJ, Wheeler KA, McIntosh N, Bass CA (1984) Temperature Variability of Very Low Birthweight Infants Nursed in Infant Servo Controlled Incubators. Chapter 3 pp13-20. Proceedings of the 2nd International Conference on Foetal and Neonatal Physiological Measurement. Oxford, 2-4 ed Rolfe P, Butterworths, London 1996.

3. Ducker DA, Lyon AJ, Ross Russel R, Bass CA, McIntosh N (1985) Incubator Temperature Control: Effects on the Very Low Birthweight Infant. Archives of Disease in Childhood 60: 902-907.

4. McIntosh N, Ducker N, Bass CA (1989) MARY - a computerised neonatal cot monitoring system. Intensive Therapy and Clinical Monitoring ; 10: 272-282.

5. Mok Q, Bass CA, Ducker DA, McIntosh N (1991) Temperature instability during nursing procedures in preterm neonates. Archives of diseaese in childhood. 66: 783-786

6. Bass CA, Badger PA., McIntosh N (1991) Mary3 - A new generation computer monitoring system. Proceedings of the 4th International Conference of fetal and neonatal monitoring pp195-197 Ed HN Lafeber, Elsevier Science publishers. Netherlands

7. Buetow KC, Klein SW (1964) Effect of maintenance of 'normal' skin temperature on survival of infants of low birthweight. Pediatrics; 34: 163-169.

8. Silverman WA, Fertig JW, Berger AP (1958) The influence of the thermal environment upon the survival of newly born premature infants. Pediatrics, 22: 876-886.

9. Day RL, Caliguiri L, Kamenski C, Ehrlich F (1964) Body temperature and survival of premature infants. Pediatrics; 34: 171-181.

10. Glass L, Silverman WA, Sinclair JC (1968) Effect of thermal environment on cold resistance and growth of small infants after the first week of life. Pediatrics; 41: 1033-1046.

11. Hey EN, Katz G (1970) The optimal thermal environment for naked babies. Archives of disease in childhood; 45: 328-334.

12. Messaritakis J, Anagnostakis D, Laskari H, Katerelos C (1990) Rectal-skin temperature difference in septicaemic newborn infants. Archives of disease in childhood; 65: 380-382.

Chapter 8

Temperature Control in Preterm Infants - Effect of Birthweight and Gestational Age.

ANDREW LYON / MADELINE PIKAAR

Introduction

The body has many mechanisms for coping with thermal stress and it will attempt to control deep core temperature within a very narrow range which is optimum for normal function and for survival [1]. The newborn infant, and in particular the immature baby, not only loses heat readily but also has difficulty in mobilising thermoregulatory mechanisms [2]. Core temperature is more variable in the neonate compared with older children or adults, and as a consequence there is an increased metabolic stress on the baby as it attempts to maintain thermal balance. It is probable that this is in part the reason for the close association between thermal stress and increased mortality and morbidity in the newborn infant [3].

By measuring core temperature we can see how effective the body is at achieving stability, but this single reading does not give any information on the metabolic cost of maintaining thermal balance. There are however no simple ways, which can be readily and continuously applied in the neonatal unit, of measuring heat loss/gain or of monitoring the metabolic effects of thermal stress on the baby.

We are left therefore with information from temperature probes as our only practical means of assessing the thermal status of the baby. However measurement of a single temperature, at infrequent intervals, is of little use in assessing what is a dynamically changing situation. One of the early responses to thermal stress is a change in vasomotor tone with peripheral vasoconstriction in the presence of cold, and vasodilatation when overheated. Continuous measurement and display of the central-peripheral temperature difference will show changes suggestive of increasing thermal stress before any significant alteration in core temperature.

There is debate about whether rectal temperatures are reliable and safe [4, 5] but, whatever one believes, it is almost impossible to retain a probe in the same place in the rectum for any length of time, and so this route is of limited use for continuous monitoring. Abdominal skin is often used as a measure of the central temperature. There is however little information on what is normal for skin temperature or for the central-peripheral difference in newborns, and in particular for the extremely low birthweight infant in intensive care.

Method

In the Simpson Memorial Maternity Pavilion, Edinburgh, preterm infants are nursed in incubators using air mode control, with humidity added (rH 80 %) for the first week of life if the gestation is 30 weeks or less. Temperature probes are placed on the abdominal skin over the liver (Tc) and on the sole of one foot (Tp). A computer monitoring system automatically collects and stores physiological data, including the two temperatures, from the monitors attached to the baby. This information is also displayed as trend graphs on the screens at the cotside [6]. The initial setting of the incubator is based on recommendations for neutral thermal environment but the nurses then alter the air temperature as necessary. They aim to keep abdominal skin temperature between 36.8 °C and 37.2 °C but they are also influenced by the central-peripheral temperature difference which they try to maintain below 1 °C.

There is now monitoring data from more than 2000 babies collected over a 4 year period with, in many cases, temperature recordings from abdomen and foot for at least the first 5 days of life. A computer programme has been developed to analyse this data and determine ranges of values for each postnatal day in preterm infants below 1500g birthweight.

Results

Effect of Birthweight and Postnatal Age

Data were obtained for the first 5 days of life from 79 babies below 1000g (with approximately 100,000 data points for each of the 5 days) and from 71 with a birthweight between 1000-1499g (approximately 80,000 data points per day).

The changes with postnatal age of Tc, Tp and temperature difference (Td) are shown in figures 8.-1 - 8.-3 for <1000g and figures 8.-4 - 8.-6 for 1000 -1499g birthweight. The values are plotted as **medians with interquartile ranges.**

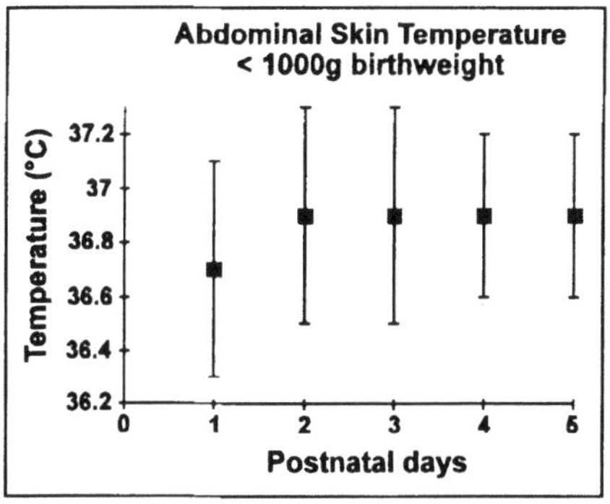

Fig. 8.-1

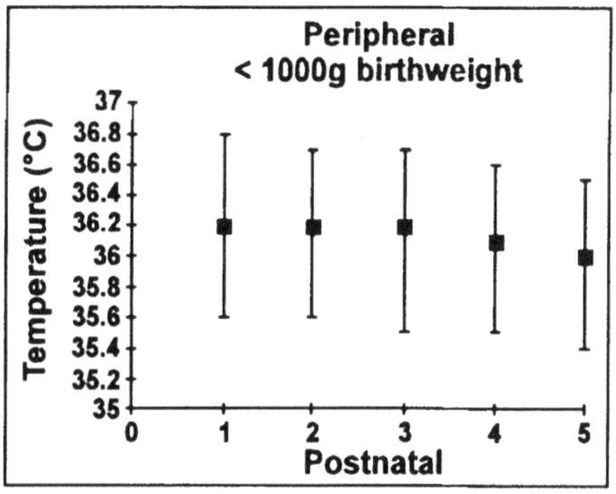

Fig. 8.-2

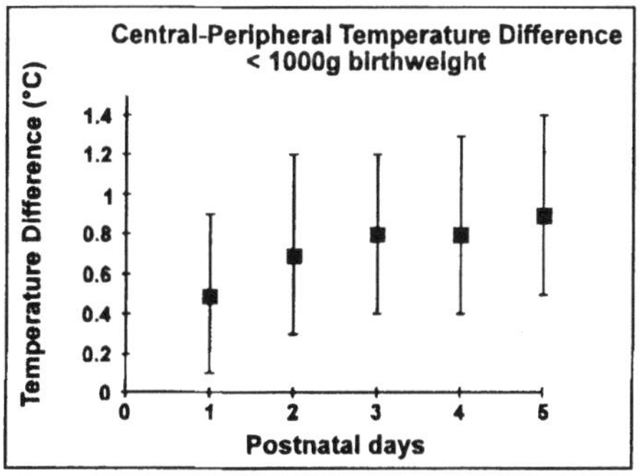

Fig. 8.-3

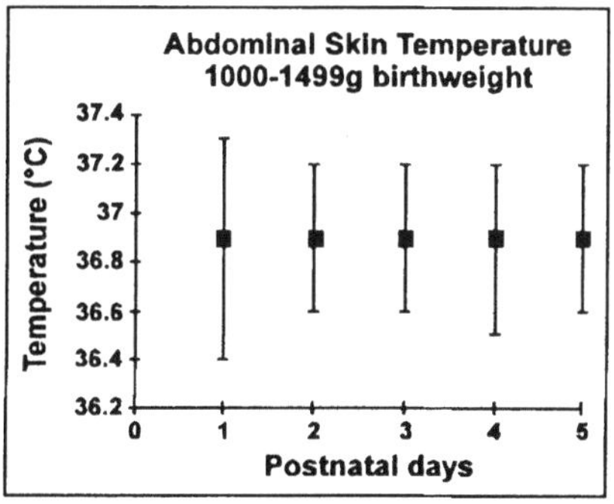

Fig. 8.-4

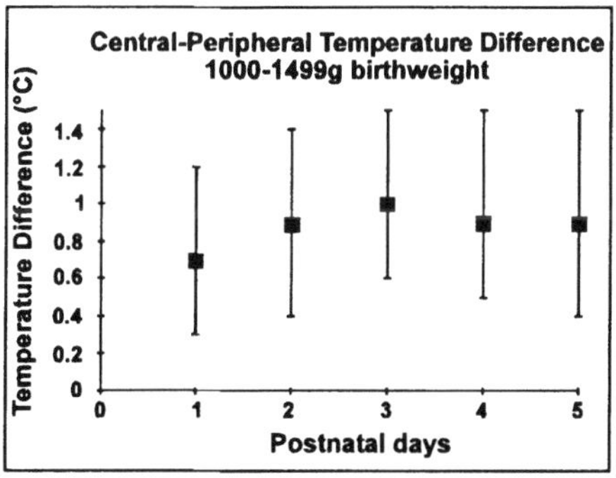

Fig. 8.-5

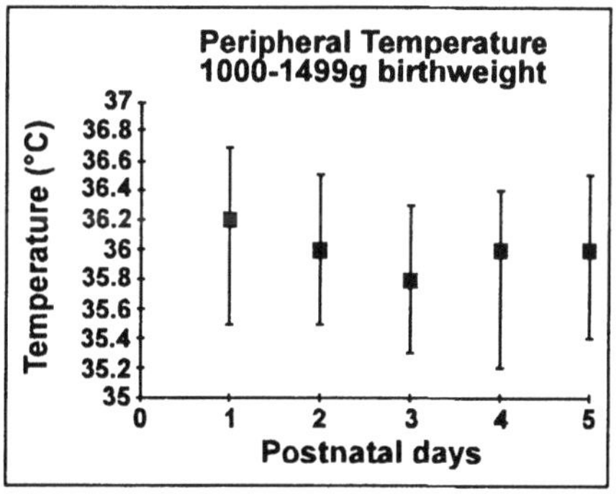

Fig. 8.-6

In the heavier babies, abdominal skin (Tc) had a constant median value of 36.9 °C on all 5 days. In babies <1000g birthweight, Tc was lower on day 1 with a median of 36.7 °C, rising to 36.9 °C on day 2 and then remaining constant at that value. Both groups had a wide range for Tc which narrowed with increasing postnatal age. Many babies had abdominal skin temperatures above 37 °C for substantial periods of time and in these cases any attempt to reduce Tc, by lowering incubator temperature, caused a rise in Td suggesting increased cold stress. In all babies in the first 2 days of life, Tc was consistently above 37.5 °C for 12 % of the total time with this proportion falling over subsequent days.

The median abdominal-peripheral temperature difference (Td) increased with time to a steady level of 0.9 °C in both groups by day 5. This value was achieved by day 2 in the heavier babies. The infants with birthweight <1000g had a lower value of Td on day 1 and this did not reach 0.9 °C until the fifth day of life. In both groups the range for Td increased with postnatal age.

A typical example of a computer graph of temperature in an infant <1000g birthweight is shown in figure 8.-7. A consistent pattern with three recognisable periods was noted

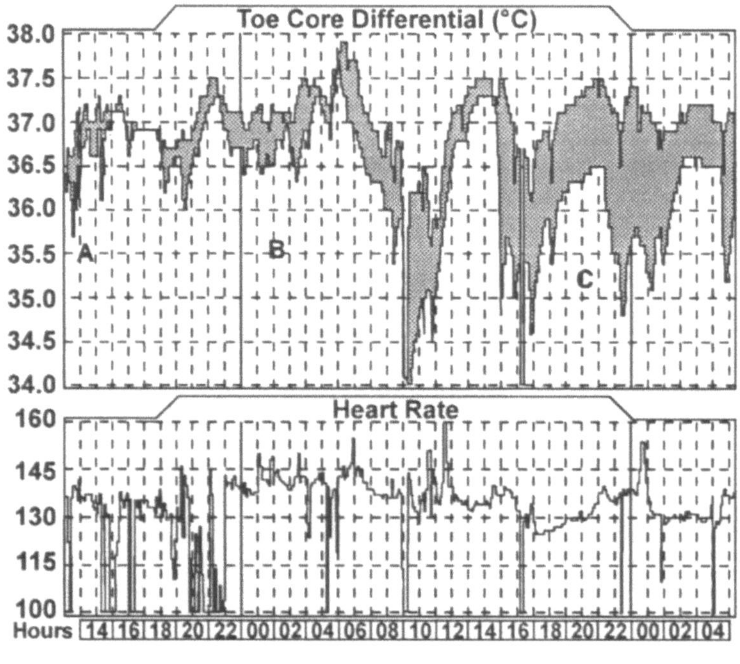

Fig. 8.-7 : Computer screen of a baby <1000g birthweight in first 2 days of life showing typical pattern of central-peripheral temperature difference.

A For several hours after birth, the abdominal skin temperature and that of the foot were very similar and tended to follow each other closely (Tc=Tp with Td always close to zero). The peripheral temperature was often higher than that of the abdomen (Tc<Tp; Td<0) and this was not due to artefact caused by poor application of the probes to the baby. Overall on day 1, Td was found to be negative (Tc<Tp) for 18.9 % of the total time.

B After a variable time, the more normal pattern with Tc>Tp was seen but there were still frequent episodes of temperature reversal, often associated with handling.

C The episodes of temperature reversal became less frequent and falls in Tc were associated with larger drops in Tp which was the expected pattern of response to cold stress. The overall result was a steady rise in the median value of Td with increasing age and a reduction in percentage of time that Td<0 - 4.9 % on day 5.

A large central-peripheral temperature difference may also be caused by poor perfusion secondary to hypovolaemic shock. In this case there should be an associated increase in heart rate and possibly some fall in blood pressure, although the latter may be a late event. Further analysis of the temperature data in the infants less than 1000g birthweight showed that when Td>2 °C there was other evidence of hypovolaemia (rising heart rate and/or falling blood pressure) for only 11 % of the time.

All these preterm babies were screened for infection and treated with antibiotics. There was no evidence of any significant association between a high Td and sepsis in these preterm babies at this stage of their life, although others have reported that this may be an indication of infection in these infants [7].

SGA vs AGA

Small for gestational age infants (SGA) with a birthweight below the 10th centile have increased problems with temperature control, when compaterd with appropriate for gestational age (AGA), because of their poor insulation. Figures 8.-8 and 8.-9 show the change with postnatal age of median Tc and Td for SGA infants and can be compared with those of AGA babies shown in figures 8.-10 and 8.-11.

The AGA babies at these gestations maintain their Tc at a constant 36.9 °C with a narrowing in the range of values with increasing postnatal age. This is associated with a rise in Td to a value around 0.9 °C by 5 days of age.

The SGA infants at the same gestations have more unstable central temperatures with a rise in the median value and a decrease in range of Tc with increasing postnatal age. Td is lower in the first 4 days of life than that of the AGA group but rises to around 0.9 °C by day 5.

At gestations above 32 weeks there were no significant differences between AGA and SGA babies.

The values are plotted as **medians with interquartile ranges.**

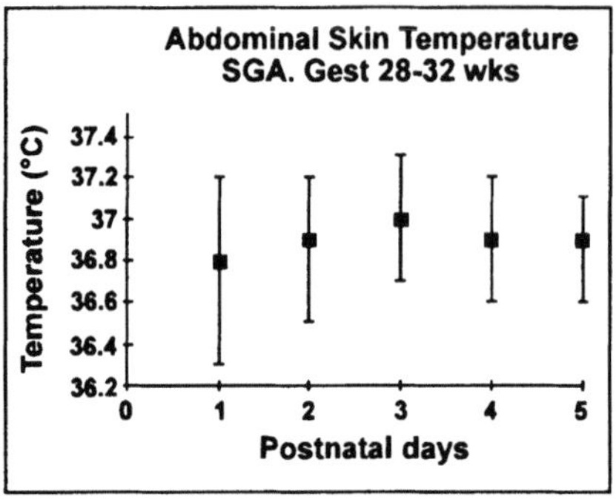

Fig. 8.-8

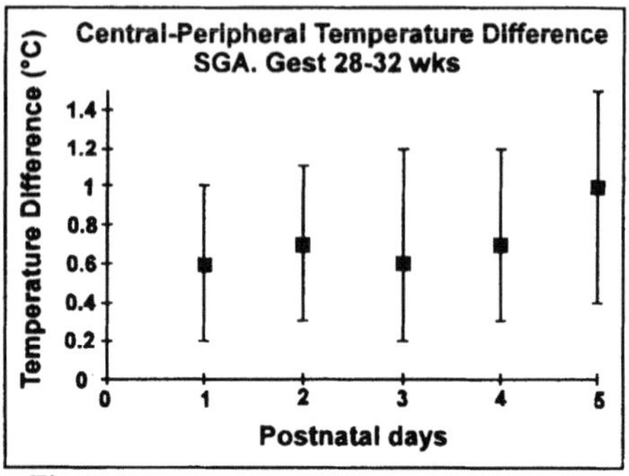

Fig.8.-9

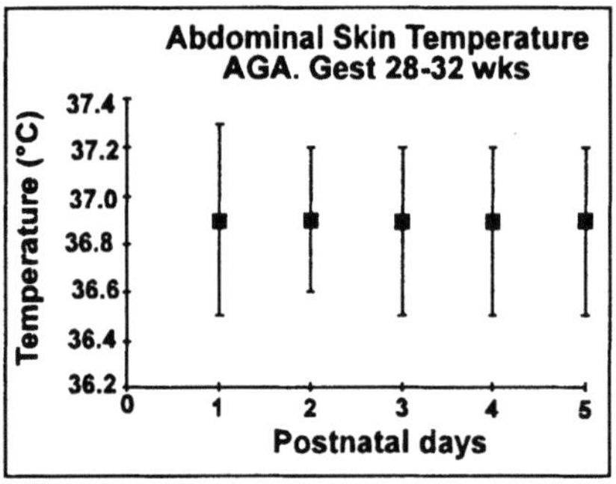

Fig. 8.-10

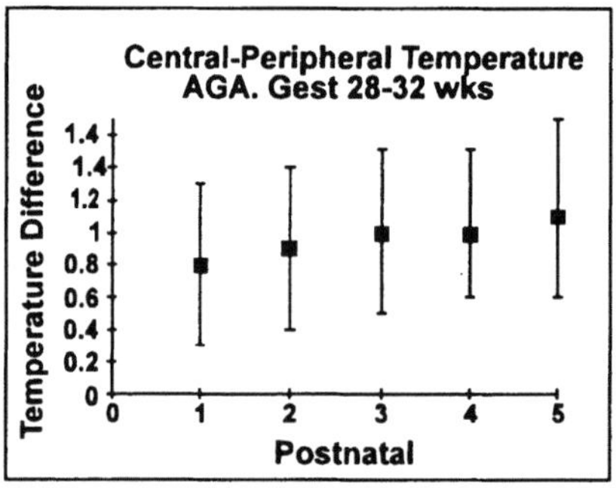

Fig. 8.-11

Discussion

These observations are derived from a very large number of data points and as such probably represent a good description of temperature changes in very immature babies undergoing intensive care. There are no data to relate them to heat flux or to measures of metabolic rate in the baby but they represent the range of temperatures at which preterm infants stabilised in intensive care when cared for by experienced nursing staff. The values give a good description of the "comfort" range for these infants. Although the nurses initially aimed to maintain Tc within a range of 36.8 °C to 37.2 °C they became convinced that following Td was a better method of determining optimum temperature control for the baby. As a result they would allow Tc to rise above 37.2 °C if Td was more than 1 °C and they felt clinically that the baby was under cold stress.

In older babies and children, exposure to cold stress will result in peripheral vasoconstriction and an increase in the central-peripheral temperature difference. In the immature babies in this study there is evidence that vasomotor tone is poorly developed at the time of birth but matures over the first 5 days. The rate of maturation is dependent on birthweight, being faster among the larger babies. These infants are unable to alter blood flow to the skin in response to thermal stress and during this time their central temperature follows closely that of the environment in which they are nursed.

Once a more normal vasomotor response was seen there were still acute episodes in which the central temperature decreased rapidly, falling below that of the periphery. This reversal of the normal pattern with Tc<Tp was possibly due to evaporative heat losses during handling. Opening the incubator doors results in a rapid fall in humidity, and water and heat are quickly lost from the exposed abdomen while the feet tend to remain covered with bubble plastic minimising evaporative losses. It would be interesting to study this effect under an open radiant warmer where there are even higher evaporative heat losses.

In this study, a rise in Td was rarely associated with other evidence of hypovolaemic shock - for nearly 90 % of time a rise in Td was due to cold stress. Similarly there was no evidence that the changes in Td were due to sepsis.

Small for gestational age infants had more problems with temperature control and showed a more immature pattern of development of vasomotor control than those born AGA. This difference between SGA and AGA was not demonstrated after 32 weeks gestation.

Implication for Practice

Nurse infants below 1500g birthweight in humidified incubators for at least 1 week.

Alter air temperature so that Td is 1 °C. This may mean that Tc rises above 37 °C.

Tc>38 °C with Td>1 °C should be further investigated to exclude sepsis or hypovolaemia.

Td<1 °C - check for signs of heat stress - usually a rising heart rate. Reduce incubator temperature.

Recommendation for Practice

The maturation of the thermoregulatory response in extremely low birthweight infants

Is there a better indicator of core temperature - is it possible to measure a zero heat flux temperature

What is the relationship between changes in Td and oxygen consumption

Is servo mode better than air mode control in incubators - which temperature should "drive" the servo - should this be Td ?

References

1. Houdas Y, Ring EF (1982) Human body temperature. Its measurement and regulation.Plenum press,New York and London.

2. Brück K (1992) Neonatal thermal regulation. In: Polin R, Fox WW eds. Fetal and neonatal physiology. Philadelphia USA, 488-515.

3. Sinclair JC (1992) Management of the thermal environment. In: Sinclair JC, Bracken MB. Effective care of the newborn infant. Oxford University Press ;40-58.

4. Merenstein GB (1970) Rectal perforation by thermometer. Lancet;I:1007

5. Morley CJ, Hewson PH, Thornton AJ, Cole TJ (1992) Axillary and rectal temperature measurements in infants. Arch Dis Child 67:122-5.

6. McIntosh N, Ducker DA, Bass CA (1989) MARY - a computerised neonatal cot monitoring system. Intensive Therapy and Clinical Monitoring 10:272-82.

7. Messaritakis J, Anagnostakis D, Laskari H, Katerelos C (1990) Rectal-skin temperature difference in septicemic newborn infants. Arch Dis Child 65:380-2.

Chapter 9

Thermal Monitoring in Sick Neonates

JOHANNES MESSARITAKIS / DIMITRIS ANAGNOSTAKIS

Thermal care of newborns and particularly low birth weight babies has become a hot topic during the past several years. This particular infant population is vulnerable to bacterial infections and because prodromal clinical signs are very unspecific, early detection of this severe disease state is of paramount importance for the prompt start of treatment [1-5].

Measurement of the boby temperature remains a reliable method to assess the thermal condition of a newborn.

This is still accomplished by means of liquid expansion thermometers or by various types of electronic thermometers or skin sensors attached to the place until the temperature becomes stable [6-8].

The reflection of real body temperature is that of the central (core) temperature, but the site of its measurement has been a point of dispute. Finally, most scientific investigations have come to the conclusion, that the rectal temperature remains as the measurement that mostly approximates a core temperature, taking into account that the real one is probably only that of the hypothalamus [6, 9, 10, 11, 12, 13].

On the other hand we must be aware of the fact, that there are factors adversely affecting rectal temperature, such as the presence of regional infection, total care nursing procedures in preterm neonates, use of hypothermia during surgery, which brings cool blood close to the rectum from cooling catheters in the legs. As a result, the rectal temperature is sometimes significantly lower or higher than the central temperature [10, 14].

Another point in evaluating the real heat condition of newborns is to define the normal range of their body temperature.

Surprisingly, there are litle data on this subject as most studies have reported on ill babies.

However, pertinent investigations are in agreement, that normal range of rectal temperature in infants is between 36.7 and 37.9 °C [7, 9, 11, 15, 16].

In our study with 55 healthy newborns, the normal rectal temperature was defined as 36.7 °C with a S.D of 0.4 °C [20]. See tab. 9.-1 and fig. 9.-1.

Common practice for thermal recording of babies placed in incubators is based, quite often, on the measurement of a single accidental temperature.

Several studies have shown that the development of significant fever is not a common phenomenon in preterm infants, and that fever, when present, is infrequently associated with systemic infection, when only a single elevated reading occurred [1, 15, 16].

Moreover, fever in an infant without other signs of infection, such as respiratory distress, apnea, lethargy, irritability, anorexia, jaundice, abdominal distension, is not common [2, 15, 17, 18]. On the other hand, the detection of high 39≥ C, persisting or recurring fever must be considered as an important index of bacterial sepsis [15, 16, 18]. The incidence of fever in fullterm newborns with sepsis ranges in various studies from 51 % to 91 % and the incidence of hypothermia is referred to 15 %. Preterm babies react to bacterial sepsis usually with normal temperature, less frequently with fever and those critically ill with hypothermia [2, 15, 17, 19, 20].

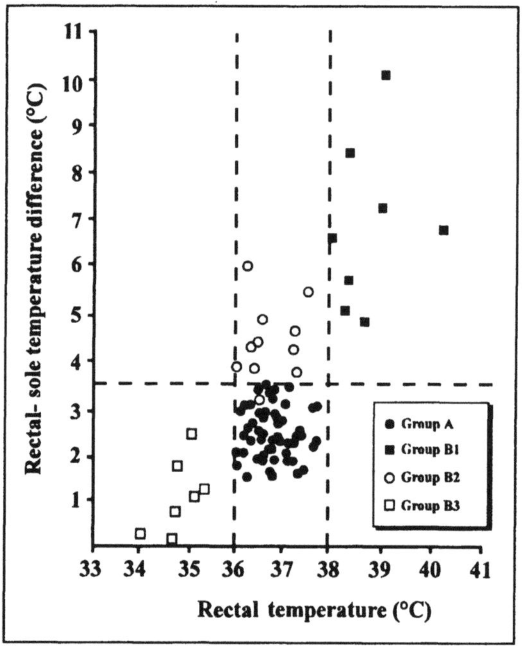

Fig. 9.-1 : Rectal temperatures and rectal-sole temperature difference in all four groups of newborn infants. The horizontal dotted line shows the critical cut point of rectal-sole temperature difference of 3-5 °C (see text). The vertical lines divide hypothermic, normothermic and hyperthermic neonates.

The clinical pediatrician in the care of newborns is frequently involved with problems of differentiating environmental from disease-related fevers or fevers of various organic etiologies and most important with problems of early diagnosis and prompt treatment of neonatal septicemia.

The correct manipulations to estimate the real thermal condition of a newborn baby and to evaluate the results has caused great concern, expecially because all the investigators tried to find another useful clinical sign in the early diagnosis of important disease states, thus avoiding risks and unnecessary investigations and treatments.

To enhance the quality and handling of temperature measurement the dual system of regular, simultaneous temperature recording or continuous monitoring of central and peripheral temperatures was introduced by few investigators [7, 14, 20, 21]. This was an attempt to provide the pediatrician with higher quality information than the measurement of a single instant temperature could provide [14, 20, 21, 22].

Thus central (core) and peripheral temperature recordings and the difference between them has considerably contributed to increase knowledge concerning the evaluation of real thermal condition and the solution of very important and serious health problems of the nursing newborn [14, 20, 21, 22]. So, measurements of paired rectal and anterior mid-lower leg skin temperatures have been applied in one group of full-term neonates with environmentally manipulated fever and in another with disease-related fevers [21].

Rectal-leg skin temperatures differences between -1 and +4F (mean+1.14±1.0) were reported in newborns whose temperature was manipulated environmentally and between +5.7 and +12.9F (mean+7.90±1.88) in infants with disease-related fevers (tab. 9.-2).

In another interesting study abdominal and toe temperatures were continuously monitored in preterm neonates during total nursing procedures. It was proved that these procedures caused large drops in both central and peripheral temperatures, with widening of the central-peripheral gap. The recovery took up to two hours, time which is enough to produce alternations in the enviromental temperature with consequent thermal stress to the infants, thus influencing their ultimate outcome [14].

Our original prospective work comprised 55 healthy and 26 septicemic infants. The diagnosis of septicemia was made on clinical grounds confirmed by positive blood or cerebrospinal fluid cultures or both in 21 infants and by necropsy findings in five. The purpose was to take serial skin (sole) and rectal temperatures recordings to find out whether septicemic newborns can exhibit any thermoregulatory reaction to their disease and whether regular temperature measurements could help in the early diagnosis of septicemia.
The 81 newborns of our study were divided in four groups. (tab. 9.-1, 9.-3)

Group A comprised 55 healthy full-term infants, group B_1 included 8 feverish full-term septicemic infants, group B_2 11 septicemic neonates, who were not feverish and in comparatively good clinical condition and group B3

7 septicemic newborns, who were hypothermic and in a poor condition, 5 of whom died. Comparability data are shown in table 9.-3. In table 9.-1 we see that all feverish septicemic infants (group B_1) had a low sole temperature (31.8 °C) and a big rectal-sole temperature difference (6.9 °C). This result was also true for group B_2, where although the rectal temperature was similar to that of the healthy babies, the sole temperature was significantly lower than that of the healthy newborns; as a consequence the temperatures difference (4.7 °C) was much larger than that of the healthy babies ($p<0.001$). In contrast the temperatures difference (1.1 °C) was low in the extremely ill babies of group B_3. In this group the sole temperature was almost the same as that of the healthy newborns, but the rectal temperature was low. Plotting our data graphically (fig 9.-1) we conclude that if the critically ill newborns of group B_3 with low rectal temperatures are excluded, only the healthy babies had a temperature difference of less than 3.5 °C. All feverish septicemic babies (group B_1) and 10 of the 11 septicemic infants with normal temperatures in group B_2 had a temperature difference of more than 3.5 °C. In other words only one of the 19 septicemic infants (5 %) of groups B_1 and B_2 had the same rectal-sole temperature difference as the healthy babies. The same temperature recordings were also applied in another group of 21 infants with 10 preterm ones among them, in whom septicemia was suspected both on clinical grounds and laboratory findings of high white cells and differential counts, but with negative blood and cerebrospinal fluid cultures. Four of them were feverish and in good clinical condition, 14 had normal temperatures and in relatively good clinical condition and 3 had normal temperatures but were in a poor clinical condition. They all reacted in the same thermoregulatory manner as their counterparts with proved septicemia. Four reacted in the same way as group B_1, 11 in the same way as B_2 and 3 in the same way as group B_3. This means that our policy of dual thermal recording is another useful and strong clinical assistance in our decision to early treatment in newborns with suspected septicemia.

Tab. 9.-1 : Mean (SD) rectal and sole temperatures and rectal-sole temperature difference in the four groups of newborn infants (given as °C)

	Rectal temperature	Sole temperature	Rectal-sole temperature difference
A	36.7 (0.4)	34.2 (0.7)	2.5 (0.5)
B1	38.7 (0.7)	31.8 (0.4)	6.9 (1.7)
B2	36.8 (0.6)	32.1 (0.9)	4.7 (0.8)
B3	34.8 (0.4)	33.7 (0.7)	1.1 (0.8)

Tab. 9.-2 : Rectal Temperature-Leg Temperature (RT-LT) Differences in Group 1 and 2 Infants (Mean ± SD)

Condition of Infant	Number of Subjects	Number of Measurements	RT-LT (F)
Normal	41	78	1.14 ±1.00
Febrile	13	13	7.90 ±1.88

Tab. 9.-3 : Comparability of the 81 newborn infants

	Group A (n=55)	Group B1 (n=8)	Group B2 (n=11)	Group B3 (n=7)
Boys/girls	31/24	5/3	6/5	4/3
Fullterm/preterm	23/32	8/0	4/7	2/5
Mean (range) birth weight (g)	2350 (1200-3850)	2420 (2710-4880)	2310 (1250-3980)	1750 (1300-3100)
Mean (range) gestational age (weeks)	36.5 (28-40)	39.5 (38-43)	36.9 (29-40)	33.0 (29-38)
Mean age at temperature measurements (days)	8	8.9	8.5	8.9

Evaluating the data of our work we can conclude that:

1. Feverish full-term septicemic infants exhibited a good thermoregulatory reaction to septicemia by increasing their heat production and by reducing their heat loss by peripheral vasoconstriction, a phenomenon analogous to feverish children and adults. These resulted in a widening of rectal-sole temperatures difference of more than 3.5 °C.

2. Full and preterm septicemic newborns with normal temperatures reacted with peripheral vasoconstriction in their effort to minimize heat loss and to preserve their core temperature within the normal range, resulting in normal rectal but low peripheral temperatures. This means that the babies of this group showed an adequate thermoregulatory reaction, but not as efficient as that which was shown by their counterparts feverish full-term septicemic infants. It is worth noting that almost two thirds of the newborns in this group were premature and our observation is challenging previous reports that preterm ill infants do not react with a normal thermoregulation. The rectal-sole temperatures difference in this group of neonates was also found to be more than 3.5 °C.

3. Critically ill and mostly preterm infants have no ability to mobilize their thermoregulatory mechanisms and are unable to conserve heat, with

consequent result in an appreciable drop in their core (rectal) temperature and a rather normal peripheral temperature. It is obvious that the rectal-skin temperatures difference (1.1 °C) is far distant from the critical cut point of 3.5 °C.

Summary

The best ways of thermal monitoring in neonates and the evaluation of the recordings have been a hot point in the management of nursing newborns for several years. A single accidental measurement of temperature has been proved unreliable.

Dual, simultaneous recording or monitoring of core and peripheral temperatures provide the pediatrician with better knowledge of the thermal condition of the newborn and is helpful in differentiating environmental from disease-related fevers, in the early diagnosis of neonatal septicemia and in the detection of thermoregulatory abnormalities during total nursing procedures.

In conclusion we recommend that the dual, simultaneous thermal recording or monitoring must be applied to all babies with a birth weight below 1250 g, to neonates in critical condition irrespective of their birth weight, and to newborns even with a single elevated recording of their temperature.

Implication for Practice

Serial, dual thermal monitoring will be helpful in diagnostic and treatment policies of neonates such as:

1. Low birth weight neonates.

2. Newborns with hyper- and hypothermia.

Recommendation for Future Research

For future research we would suggest the estimation of the length of time which is required to reset the rectal-sole temperatures difference to normal range and to find out if the time which is required for this restoration has anything to do either with the seriousness of the disease or the ability to reaction of the affected newborn.
Because we believe that rectal temperature is reliable we would welcome incubators adequately equipped with facilities for monitoring skin and rectal temperatures even for short periods of time and intermittently.

References

1. Kluger MJ (1992) Fever revisited. Pediatrics 90: 846-50.

2. Kovatch AL, Wald ER (1985) Evaluation of the febrile neonate. SemPerinatol 9: 12 - 19.

3. Lieu TA, Schwartz SJ, Jaffe DM, Fleisher GR (1991) Strategies for diagnosis and treatment of children at risk for occult bacteremia: Clinical effectiveness and cost-effectiveness. J Pediatr 118:21-29.

4. Baskin MN, O' Rourke EJ, Fleisher GR (1992) Outpatient treatment of febrile infants 28 to 89 days of age with intramuscular administration of ceftriaxone. J Pediatr 120: 22-27.

5. Down SM, Mc Nutt RA, Margolis PA (1991) Management of infants at risk for occult bacteremia: A decision analysis. J Pediatr 118: 11-20.

6. Ogren JM (1990) The inaccuracy of axillary temperatures measured with an electronic thermometer. Am J Dis Child 144: 109-111.

7. Morley CJ, Hewson PH, Thornton AJ, Cole TJ (1992) Axillary and rectal temperature measurements in infants. Arch Dis Child 67: 122-125.

8. Childs C, Little RA (1994) Acute changes in oxygen consumption and body temperature after burn injury. Arch Dis Child 71: 31-34.

9. Herzog LW, Coyne LJ (1993) What is fever? Normal temperatures in infants less than 3 months old. Clin Pediatic 32: 142-146.

10. Brown PJ, Christmas BF, Ford RP (1992) Taking an infant's temperature: Axillary or rectal thermometer? NZ Med J 105: 309-311.

11. Keeley D (1992) Taking infants' temperatures. Forget the axilla- the rectum is better. BMJ 304: 931-932.

12. Anagnostakis D, Matsaniotis N, Grafakos S, Sarafidou E (1993) Rectal - axillary temperature difference in febrile and afebrile infants and children. Clin Pediatr 32: 268-272.

13. Keusch GT (1976) Fever. To be or not to be. NY State J Med 76: 1998-2001.

14. Mok Q, Bass CA, Ducker DA, McIntosh N (1991) Temperature instability during nursing procedures in preterm neonates.Arch Dis Child 66: 783-786.

15. Voora S, Srinivasan G, Lilien L, Tsu EY, Pildes R (1982) Fever in full-term newborns in the first four days of life. Pediatrics 69: 40-44.

16. Osborn LM, Bolus R (1985) Temperature and fever in the full-term newborn. J Fam Pract 20: 261-263.

17. Klein JO, Marcy SM (1990) Bacterial sepsis and meningitis. In: Remington JS and Klein JO, Eds, Infectious diseases of the fetus and the newborn infant. WB Saunders pp 601-644.

18. Crain EF, Shelov SP (1982) Febrile infants. Predictors of bacteremia. J Pediatr 101: 686-689.

19. Rutter N (1986) Temperature control and its disorders. In: Robertson NR Ed, Textbook of neonatology, Churchill-Livingstone pp 148-161.

20. Messaritakis J, Anagnostakis D, Laskari H, Katerelos C (1990) Rectal-skin temperature difference in septicemic newborn infants. Ach Dis Child 65: 380-382.

21. Pomerance JJ, Brand RJ, Meredith JN (1981) Differentiating environmental from disease-related fevers in the term newborn. Pediatrics 67: 485-488.

22. Jackson JA, Petersen SA, Wailoo MP (1994) Body temperature changes before minor illness in infants. Arch Dis Child 71: 80-83.

Optimal Thermal Environment

Chapter 10

Physical Properties of the Thermal Environment

JOCHIM KOCH

Introduction

For more than 100 years specific equipment has been used to keep newborns warm [1]. The first air-controlled incubator with integrated humidification was designed in 1947 [1] by Chapple. Since then, and especially in the last 15 years, birth weights have decreased to as low as 450g and gestational ages to as early as the 24th week of gestation. These very small newborns have different heat losses than full term babies: the heat loss by evaporation is much higher [2], the ratio of surface area to body weight is less favourable and body heat production is very limited. This results in greater requirements from warming therapy equipment and these are described in this paper.

Heat Balance

In intensive care units babies are often nursed naked for easier handling and observation. This would normally result in increased heat loss. Ambient conditions are usually too cold for a newborn baby to maintain heat balance. Therefore these babies have to be protected by a warm environment. To avoid any heat loss from the baby the immediate environment, the "micro-climate", must be controlled in such a way that the baby's own heat production and all heat losses are in balance. The "micro-climate" can be controlled with special warming therapy equipment such as incubators, open radiant warmers, heated cots etc.

The incubator air temperature at which the baby is normally in heat balance is called the 'neutral temperature' and the smaller the baby, the higher this value will be.

Each human being has a heat production which is dependent on body weight [3] with the smaller babies producing the least heat.

Heat balance can be calculated [4] and graphically represented as shown in figure 10.-1.

In the last decade an increasing number of very small babies have survived and the heat production of these infants is limited compared with those at term [20]. The modes of heat loss are also different. The comparison of heat balance in babies of different gestations is shown in figure 10.-2.

Authors have tried to find an "operative temperature" [32] by considering the air temperature of the incubator in combination with the influence of radiant heat loss to the single wall of the hood. These days all intensive care incubators used for preterm and sick neonates have double walls and this has significantly reduced radiant heat loss. Others have discussed the importance of humidity in maintaining heat balance. The effect of humidity was not considered in the concept of the "operative temperature" with the result that these calculations cannot be applied to the controlled environment produced within modern incubators. However a new computerized program has been developed [4] which can calculate the heat balance of preterm neonates and which takes into consideration all the physical properties and physiological factors shown by clinical research to play a significant role. This programme was used for the calculations in this report.

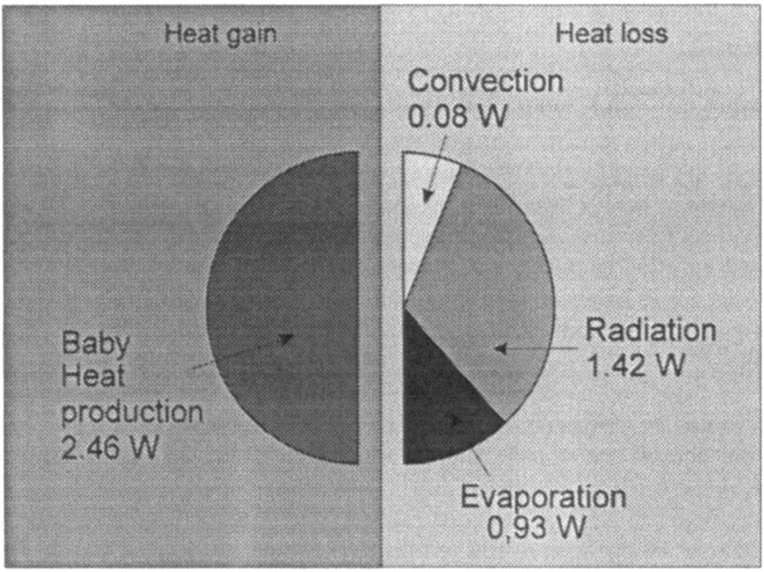

Fig. 10.-1 : Heat balance of a baby of body weight 1500g and 30 weeks gestational age (G.A.:). Approximated on the first day in an incubator with incubator air temperature at 35.8 °C and 75 % rel. humidity. The baby's heat loss and heat production are in balance. The main heat loss is radiant heat loss. 1Wh = 0,860Kcal = 1 Joule

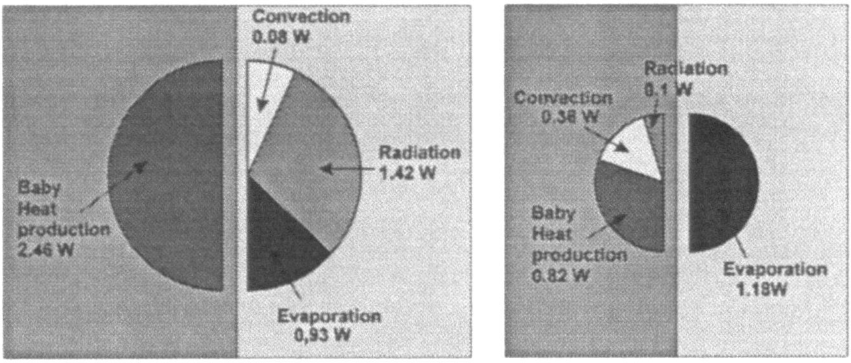

Fig. 10.-2 : Heat balance of a baby (the same baby is represented in Fig 10.-1) of body weight 1500g and 30 weeks G.A. (on the left) compared with the heat balance of a baby of body weight 500g and 24 weeks G.A. (on the right) both on the first day in an incubator. The heat production and heat loss of both babies is in balance. The heat production of the smaller baby would not normally be sufficient to maintain the baby in heat balance. The limited heat production must be assisted by providing additional convective and radiant heat thereby compensating for the very high evaporative heat loss. 1Wh = 0,860Kcal = 1 Joule

Comparing the 1500g baby of 30 weeks G.A. with that of the 500g baby of 24 weeks G.A., evaporation is the main method of heat loss in the smaller baby while in the larger it accounts for only one third of all the heat lost.

Equipment used to provide Warming Therapy.

There are four different modes of heat transfer that are important when considering warming therapy equipment. They are: conduction, convection, evaporation and radiation. Each mode of heat transfer has characteristic physical limitations which are shown in figure 10.-3.

a) Heat transfer to the baby by **conduction** is dependent on the surface temperature of the mattress and, for conductive heat transfer, this is limited 39 °C [5] to avoid the risk of skin burns.

b) Heat transfer to the baby by **convection** is dependent on the air temperature of the incubator and this is limited to 39 °C [6]. Although the air temperature in itself will not cause burns, surfaces and objects surrounding the baby will have the same temperature as the air and could burn skin by conductive heat transfer.

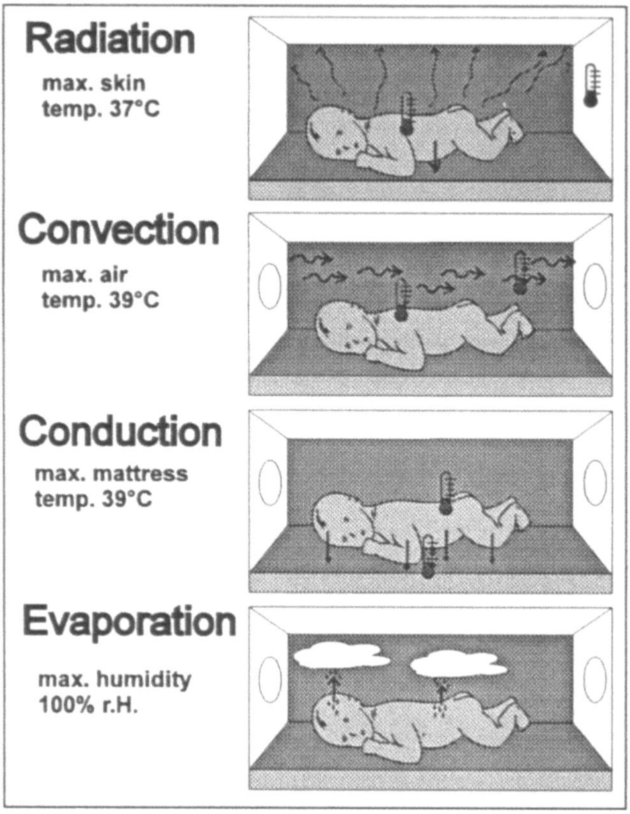

Fig. 10.-3 : This graph shows the limitation of heat transfer to the baby by the four different modes: conduction, convection, radiation and evaporation. It can be seen that each mode of heat transfer has unique limitations given by physical or physiological restrictions. If the baby needs more heat than can be transferred by one mode, it must be combined with one or more other modes in order to increase heat transfer.

c) Heat transfer to the baby by **radiation** is limited so that the maximum radiant heat power transmitted to the baby does not exceed 600 W/m^2 [7]. With radiant heat there is a constant risk of skin burns [7] and therefore equipment emitting radiant heat to the baby should include skin temperature control and/or a time limit control [7, 8].

d) Heat transfer by **evaporation** is always a heat loss from the baby and can be limited by providing the maximum relative humidity possible (100 % r.H.). However providing relative humidity of 100 % results in water condensation on the hood of the incubator obstructing visibility of the baby. Therefore most hospitals limit incubator relative humidity to a just below the 'dew point' (70 to 85 % r.H.)

Hospital equipment used in warming therapy in the paediatric field provides either one mode of heat transfer (e.g. Open radiant warmers) or a combination of these modes (e.g. Incubators). This equipment is described below.

Incubators

Incubators can be described as a transparent closed housing with an inner air conditioned "micro-climate" providing a higher air temperature than the ambient environment. The incubator's closed design allows a higher humidity and oxygen concentration than the ambience. They should comply with the International Standards for Incubators [6]. The maximum heat transfer to the baby in an incubator is normally not more than 100 W/m^2 as opposed to radiant warmers which can provide up to 600 W/m^2 [8]. This factor explains the limited ability of an incubator to warm up a hypothermic baby.

One advantage of the incubator is that it provides protection against bacterial infection due to the design principle of a closed hood and filtered ambient air. On the other hand the incubator creates a physical barrier between the baby and the nurse or parents.

Convection

Convection is the main mode of heat transfer used by incubators to transfer heat to the baby. When the incubator air temperature is cooler than the baby's skin temperature, heat loss from the baby is achieved by the high velocity of air flow. Therefore modern incubators should have a minimal air flow velocity (6 to 8 cm/s) [26].

However a low air flow velocity reduces the ability to effectively warm up hypothermic babies within the shortest possible time.

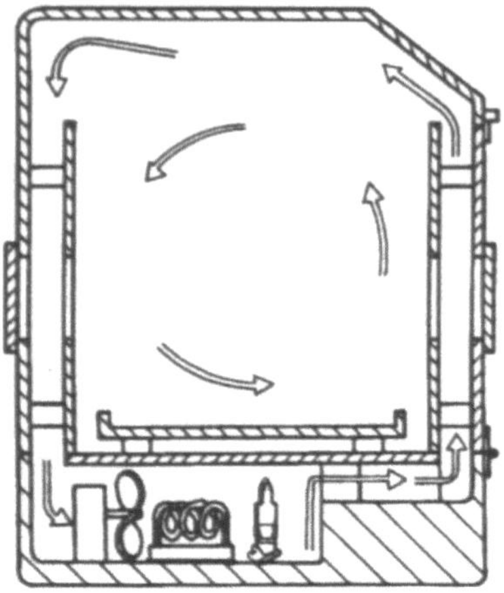

Fig. 10.-4 : A double walled incubator. Most of the air flow is channelled between the double walls reducing the air flow velocity directly above the mattress.

In incubators the air temperature is controlled by a thermostat. To prevent unacceptably high temperatures due to malfunction of this part a second, independent safety thermostat is installed [6]. The incubator air temperature can be set by the user within the normal range of 28 to 39 °C. The user can choose manual control of the incubator air temperature or the baby's skin temperature control (see Servo Control). The philosophy behind controlling air temperature is that of providing a constant environment in which the baby's heat production and heat losses caused by environmental conditions reach a state of balance. However in actuality neither the baby's heat production nor the environmental conditions in the incubator are constant. Therefore skin servo control could provide a better therapeutic alternative for treating babies in incubators.

Whilst nursing a baby in an incubator the **port holes** are often opened [27]. These disturbances can happen more than 30 times a day. However this should not result in any drop in the incubator air temperature. Modern incubators are designed in such a way that the environment inside the incubator remains more or less stable even during routine opening of the port holes.

Normally the large **front door** of the incubator is kept closed. However in order to gain better access when performing certain procedures, it may be necessary to have the front door open e.g. during intubation or weighing. In this situation the temperature inside the incubator could drop markedly [14]. The different incubators available on the market vary in their response to this situation. The baby can experience a 'cold stress' in this situation. Some incubators offer an integrated weighing scale and therefore avoid the necessity of opening the front door to weigh the baby, minimising heat loss. Preterm and low birthweight neonates need the highest incubator air temperatures because of their limited ability for heat production and high heat loss [6].

The incubator air temperature limit of 39 °C [6] could be insufficient, in some circumstances, to keep small neonates warm. This may be the case especially when the air flow velocity in the incubator is too high, radiant heat loss is increased (single walled incubator) and the relative humidity is too low. For this group of very small neonates, modern incubators optimising all of the above mentioned features should be used.

Humidity

For preterm infants with a gestational age less than 30 weeks the influence of heat loss by evaporation is highest during the first week of life [2]. This evaporative heat loss should be minimized by providing a higher humidity inside the incubator. Low incubator humidity levels will not only cause increased heat loss, they will also produce very dry skin and can reduce skin development and resistance to fungal infections.

It is known that a comfortable room humidity level for adults, especially with regard to respiration and to avoid dryness in the nose and throat, is between 40 to 60 % r.H. Why should this not also be valid for preterm babies? It must be recognized that the humidity in an incubator is greatly reduced by heating the incubator air. For example: an air relative humidity of 40 % in a hospital with a room temperature of 24 °C will be reduced to only 19 % r.H. in an incubator when the air temperature is increased to 37 °C. Even the thickened mature skin of adults will become very dry and rough under these conditions!

Heat loss by evaporation occurs mainly via insensible water loss through the skin. The more preterm the baby, the more water loss occurs through the skin [2]. The rate of evaporation is dependant on the humidity of the environment and the air flow velocity. Each ml of water that evaporates through the baby's skin consumes 0.65W (560cal) energy.

Heat loss by evaporation can be reduced by humidification of the environment. This is possible using incubators incorporating an integrated humidifier. The risk of infection must however be considered. Humidification in incubators was not accepted for many years in some countries for fear of the risk of infection [9]. If the humidification system is safe and avoids the risk of infection then there is no reason not to use humidification. This reason for not using humidity is no longer valid due to the new humidifier concept incorporated into the Dräger Incubator 8000 [10].

Higher humidity levels allow the use of lower air temperatures. As incubator air temperature is limited to 39 °C, very low birthweight neonates in the first few days of life require increased levels of humidity in order to maintain enough heat to maintain their heat balance.

Humidity is even more important if the baby is not ventilated because ventilated babies receive humidified inspiratory gases while non ventilated babies do not inspire humidified air. Modern incubators with integrated humidifiers can provide a maximum level of humidity of 85 % r.H. (see fig. 10.-6).

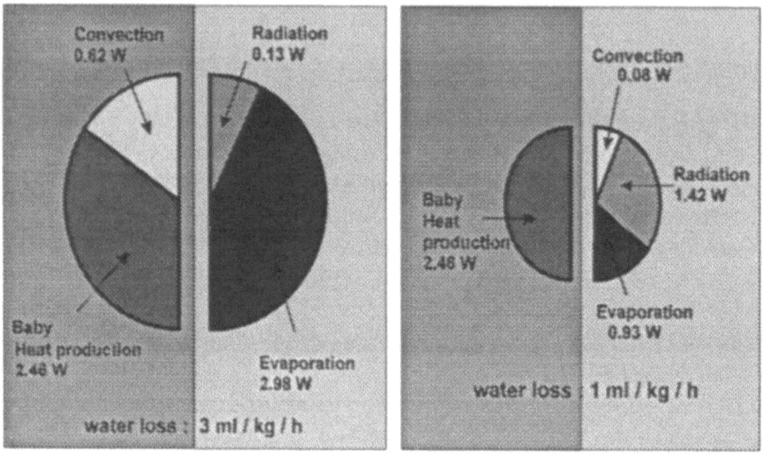

Fig 10.-5 : Heat balance of a baby of body weight 1500g and 30 weeks G.A.: in an incubator with 20 % r.H. (left) and in an incubator with 75 % r.H.(right). It can be seen that the baby in the dry incubator has double the heat loss compared to the baby in the humidified incubator. As a result the fluid turnover of the baby in the dry incubator is doubled. The air temperature can be reduced by a higher humidity. 1Wh = 0,860Kcal = 1 Joule

It has been well established that the best humidification systems produce humidity as a steam from a source of boiled sterile water that is not contaminated with condensate or aerosol. This active system keeps itself sterile because the steam and water temperature remain at nearly 100 °C. The water source does not come into contact with the air circulated to the baby. The humidity is controlled by a sensor and its level can be set by the user. Recommended values for setting humidity levels are available [11].

During neonatal intensive care the nurse handles the baby with the result that the port flaps are often open. Clinical studies show [12] that during these periods there is a drop in humidity which can result in a decrease in the baby's peripheral temperature and increased heat loss.

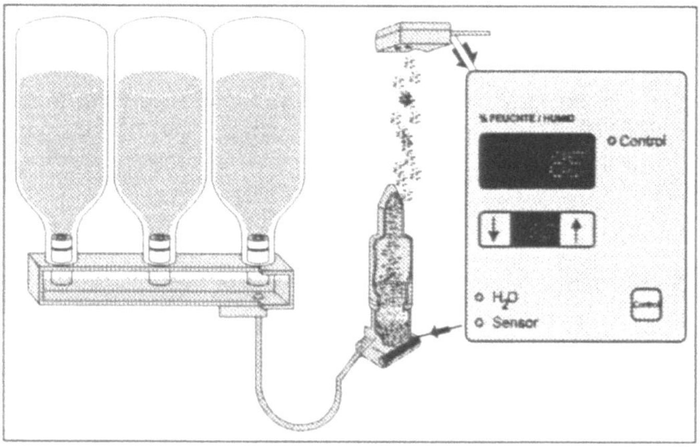

Fig. 10.-6 : Modern concept of an integrated humidifier

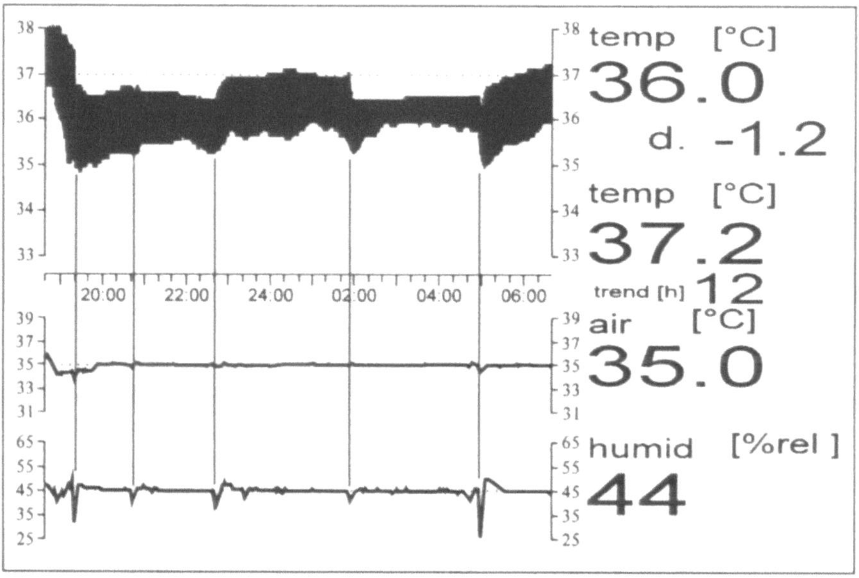

Fig. 10.-7.: Graph showing the peripheral temperature of a preterm baby in an incubator with a relative humidity of 45 % and an air temperature of 35 °C. During the time in which the port flaps are open, the relative humidity drops from 45 % to 40 %. At the same time the peripheral temperature measured on the baby's foot drops from 36 to 35 °C
1Wh = 0,860Kcal = 1 Joule

Heat loss by evaporation can be very important in preterm babies during the first days of life and therefore humidification should be controlled automatically. The humidity sensor detects a drop in humidity and reacts by increasing humidification. After opening the large front door, the time required to re-achieve the set humidity level can be shortened by 1/3rd by using this type of automatic control system [13].

The conditions within the incubator should remain as stable as possible. Modern incubators incorporating servo control of the three important environmental parameters temperature, humidity and oxygen can achieve a high degree of stability.

Radiant Heat

Heat transfer to a given surface is defined by the temperature difference between the two surfaces:

(-) A warm body surface of the baby in an incubator loses heat to the colder walls.

(+) The warm heat source of a radiant warmer transfers heat to the skin of the baby.

Heat loss by radiation can be reduced by using double walls (fig. 10.-4). The inner wall is warmer providing a buffer between the baby and the cooler outer wall which is in contact with the ambient air. Double walls prevent the loss of heat caused by cooler room temperatures [21].

Some incubators heat an area of the hood, and this is offered either as an option or is fully integrated into the design. This concept was developed many years ago with the first air conditioned incubators [15] but clinical studies have shown it to be inefficient [16].

The amount of heat radiated by a hood warmer is very small because the surface temperature of the incubator hood cannot be more than 40 °C [6]. The more heat that is transferred to the baby by radiation, the more the air temperature must be reduced to maintain the heat balance of the baby. The result is a higher convective heat loss and there is no overall difference in the heat balance of the baby, only in the contribution of the different modes of heat transfer.

The conditions in the room and any additional medical equipment can influence the heat balance causing a gain or reduction in radiant heat:

Heat Gain:	**Heat loss:**
sunlight	cold room walls
phototherapy	windows of NICU

Conduction

Heat loss by conduction is normally negligible in an incubator when a foam mattress is used. There have been reports published on the use of a heated mattress in incubators [17, 28].

Care should be taken when using a fluid or gel filled mattresses due to the high heat capacity of such equipment. If there is any temperature difference between the fluid in the mattress and the core temperature of the baby, large amounts of heat will transfer either to or from the baby. This heat transfer is uncontrolled and should therefore be avoided.

Oxygen Therapy

All incubators for babies should have the provision for administering oxygen to the inspired air in line with current standards [6]. This is often administered by a flow-meter and monitored using an oxygen analyser. These devices require constant checking and adjustment by the nursing staff. When hand ports are open the oxygen concentration drops and may take a long time to return to the set value (up to 10-14 minutes [13]) and this could result in the baby receiving insufficient oxygen for considerable periods of time.

Incubators are available with an integrated oxygen-controller which supplies a nearly constant oxygen concentration even with open hand ports (for example with open hand ports the oxygen concentration decreases by only 2 Vol. % from 40 Vol. % to 38 Vol. %) [13].

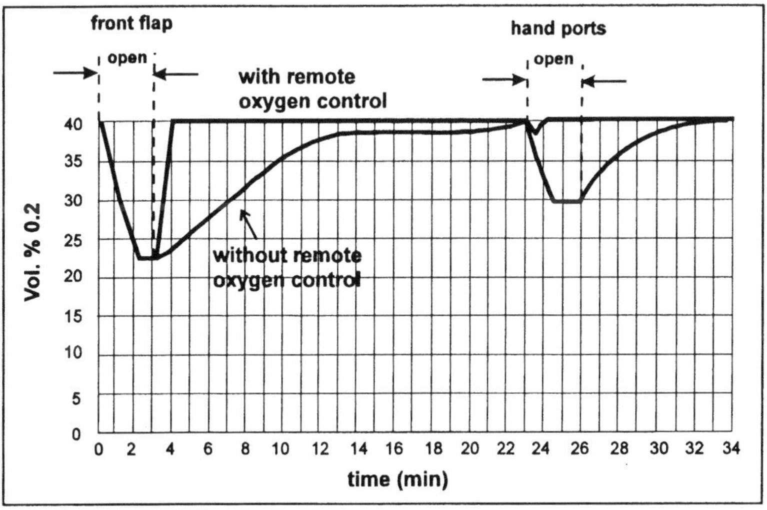

Fig 10.-8 : Graph showing oxygen concentration in the incubator hood of an incubator with and without an automatic oxygen controller. The incubator with oxygen controller returns to the set oxygen concentration within 3 minutes after closing the front flap. The incubator without an oxygen controller requires an additional 14 minutes before returning to the set oxygen concentration.

Hats

The relatively large surface area of the head of a newborn is a major source of heat loss when nursed in an incubator in which the air temperature is less than the baby's core temperature. This heat loss can be reduced significantly by applying a three-layered hat made of wool and gauze to the baby's head [21] and use of such a hat is recommended for all infants who would benefit from a controlled thermal environment [21].

Heat Shields

Previously, in single walled incubators, heat shields were sometimes used to help overcome the inability of the incubator to warm preterm babies. When placed over the baby this shield is warmed to nearly the same temperature as the surrounding incubator air and so reduces radiant and convective heat losses [28].

These shields disturb nursing care and should no longer be necessary in modern intensive care incubators incorporating double walls, low air flow velocity and active humidification systems.

Open Intensive Care

Open Radiant Warmers

Open radiant warmers are used with the heater situated directly above the baby. For the sake of better visibility and increased access for managing infusions, sensors, catheters etc the baby is exposed to cool, dry, draughty air. The radiant warmer is an advantage during resuscitation or initial intensive care because it offers better access to the baby as it has no hood or doors. It may also positively contribute to the bonding between parents and their baby. In contrast to incubators, open radiant warmers provide a constant heat transfer to the baby, even with the flaps open. The power of the heater is normally managed by servo control of the skin temperature [7, 8].

Radiant warmers can sometimes be used for short periods of time in a manual mode without servo control of the skin temperature. This mode should never be used for continuous thermal therapy for infants. The manual control mode lacks feedback information on body temperature and therefore holds a great risk of over- or under heating. If manual mode is used for short term warming the infant's condition and temperature should be checked at least every 15 minutes [7].

Under a warmer, radiant heat loss is noticeably reduced and there is usually a net gain, whereas convective and evaporative losses are considerably increased [22, 23]. When the infants are nursed under radiant warmers for more than a short term period, the increased insensible water loss must be considered in their management.

Under open radiant warmers the skin of the baby is in direct contact with room air. No humidification is generally available and the skin is at a higher temperature than that of a baby nursed in an incubator. Both these reasons are responsible for the high evaporative loss found with open radiant warmers.

The distribution of heat over the mattress is very uneven [30] and this will be aggravated by the tilted mattress tray.

Heat shields should not be used with open radiant warmers to reduce convective and evaporative heat losses because they will shield the baby from the heater [30] and the material normally used cannot transmit radiant heat.

Figure 10.-9 shows a comparison between an incubator and an open radiant warmer:

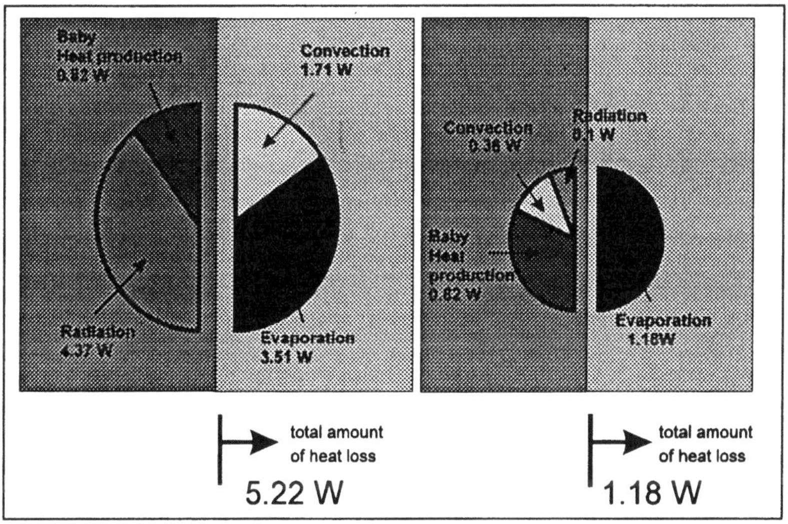

Fig. 10.-9 : Comparison of the heat balance between an incubator and an open radiant warmer. It shows that the heat loss of an open radiant warmer by evaporation and convection is much higher than in an incubator. It can however, be compared with a higher radiant heat gain. In this calculation the water loss of the baby in the incubator is 4 ml/kg/h instead of 11 ml/kg/h in an open radiant warmer. 1Wh = 0,860Kcal = 1 Joule

Other authors have shown no difference in heat balance when comparing incubators and open radiant warmers [31]. This is different to the calculation above because, although there seems to be no difference in the total heat production, there is an increased metabolic rate under open radiant warmers due to the high rate of evaporation and convective heat loss which is balanced by a higher radiative heat gain. This has to be confirmed by further studies.

The maximum heat transfer to the baby is 600 W/m^2 [8] as opposed to incubators in which it is approximately 100 W/m^2. However, the heat loss from the baby is several times more (in this example 4.5) than in an incubator and the water turnover is 2.5 times (in this example) higher. As a consequence the open radiant warmer is able to warm babies more quickly but the heat losses from the baby are also much higher and overall more fluctuant.

Some nurses have shielded the babies with a separate foil or artificial skin [18] to avoid insensible water loss during the first days of life and reduce the risk of hypothermia under open radiant warmer. However what is happening under this foil is unknown.

- No clear view of skin and complexion
- The wrap must be sterile since it touches the skin of the premature baby.
- Underneath the plastic wrap the relative humidity will be approx. 100 % causing water drops to form directly on the skin. This is a good nutrient medium for bacteria introduced from outside and may cause infection. Fungi will rapidly develop underneath the wrap.
- The wrap must be removed whenever the patient is treated and be replaced afterwards. Sterility is imperative.
- When the wrap is removed for therapy the wet body of the premature baby will be exposed to the dry environment. The relative humidity then drops from 100 % on the skin to 5 % - 20 % (i.e. by > 80 %!). The water on the skin will evaporate immediately causing the baby significant cold stress as a result of the loss of the heat of evaporation. In intensive care this is done several times a day.
- An infant who is not ventilated will breath the dry incubator air which may cause the nasopharyngeal space to dry, which in turn may result in additional respiratory water and heat loss.

Which temperature is established? There is no air temperature control. It seems to be a "substandard" incubator.

Oxygen hoods are often used on open radiant warmers (see below -Oxygen Hoods) if the baby needs a higher oxygen concentration in the inspired air.

Heated mattresses

Heated mattresses, heated cribs or heated cots are used to care for the infant once he is dressed. Cot nursing is normally inexpensive and clinically useful when close, continuous observation is not required. Heated mattresses are also used in combination with open radiant warmers to increase the heat transfer to the baby, especially in hypothermia.

Heated mattresses are also used as an alternative method of keeping small babies warm and are suitable for use in developing countries [19].

Heat can be transferred by conduction between the surface of the body of the baby and the mattress. The direction of the heat transfer (heat loss or heat gain of the patient) is dependent on the difference between the average temperature of the mattress and the average contact temperature of the baby.

For a baby lying on a foam mattress, the heat loss is given by the heat transfer of the material of the mattress and the difference between the temperature of the baby and that of the mattress underneath. Foam mattresses have a high insulation effect and heat loss is reduced to an insignificant amount. This is normally the case in incubators.

Heated foam mattresses are used occasionally for term babies. But as shown later on, this mattress cannot transfer much heat to the baby and is not very effective. Its use is restricted to babies who are covered with blankets.

A baby lying on a heated mattress will be warmed up at a rate dependent on the temperature difference of the average contact temperature and that of the surface of the mattress. The surface temperature depends on the conductance of the mattress. Heat transfer through a foam mattress is limited, but that through a gel or a liquid mattress is much higher. The surface temperature of the mattresses must be limited, because of the risk of burned skin [5], to a maximum of 39 °C. The heat transfer to the baby is limited by the maximum temperature of the mattress and the average contact temperature of the baby. When the baby has the same average contact temperature as the mattress the heat transfer will be reduced to zero! As a consequence a heated mattress can only transfer heat to the baby (or from the baby) if there is a temperature difference between the baby and mattress (see fig. 10.-10). A heated mattress with a heat transfer of more than 230 W/m^2 should have mattress temperature continually monitored [5] to avoid risk of burned skin. Heated mattresses work in a normal range of heat transfer rates of 5 (low heat transfer) to 600 W/m^2.

If the mattress temperature is set to the normal range of the average contact temperature (37-38 °C) it acts like an automatic temperature control system. If, however, the mattress temperature differs from the normal range of the average contact temperature, there are some problems which have to be considered. For example, if a fluid filled mattress is 2 °C colder than the average contact temperature, there is a heat loss from the baby as long as both temperatures are not in balance. The same occurs to a mattress which is 2 °C warmer than the average contact temperature of the baby. In this case the baby will be warmed up as long as the mattress is warmer than the baby's average contact temperature. It is important that heated mattresses should be controlled by their mattress temperatures in accordance with published standards [5].

If there is a high heat transfer between mattress and the core (for example: fluid filled mattresses, gel mattresses etc.) it is important to remember that the core temperature of a small baby will finally equilibrate with the mattress temperature. The consequence is that such heated mattresses should only be set to the normal core temperature between 37 and 38 °C. When the baby needs to be warmed up, the mattress temperature can be set to a maximum of 39 °C.

Caution: If the temperature of a mattress with a high conductance is lower than the average contact temperature of the baby, the baby loses heat!

The mattress temperature should be set lower than 37 °C only when the baby shall be cooled down.

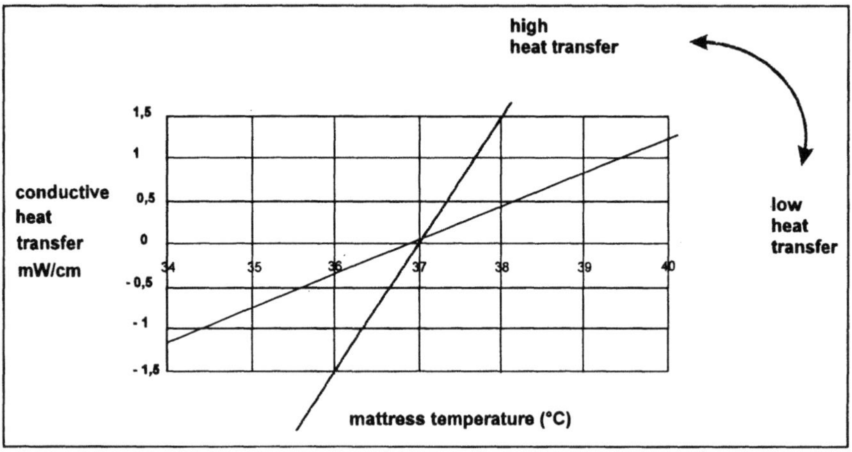

Fig. 10.-10 :Heat transfer of a heated mattress to the baby is dependent on the temperature difference between core and mattress. It shows, that the heat transfer to the baby will be reduced to zero when core and mattress have the same temperatures. The consequence is, that heated mattresses can only transfer heat to the baby (heating or cooling!) if there is a temperature difference.

Heated mattresses can be used instead of incubators for term babies [19]. The advantages are the simplicity and the easier access to the baby.

Caution should be given when using heated mattresses in incubators if it has not been approved by the manufacturer: In this case there are two different heat sources with two different servo control systems which could influence each other and possibly result in too high a temperature in the baby.

Combined Use

During resuscitation babies are often covered with blankets. In this case the radiant heater cannot warm the baby and heated mattresses are often used to compensate for lack of radiant heat.

Heated mattresses are used in combination with open radiant warmers as an additional heat source and for more constant, comfortable conditions. Clinical results show that the fluctuations in the power of the radiant warmer are reduced significantly [33], and this is associated with a reduction in water loss.

Transport Incubators

In special circumstances small preterm infants have to be transported to a neonatal intensive care unit. In this case, all requirements of an incubator have to be fulfilled in a special transportable incubator. Additionally, this incubator needs its own supply of energy (battery for electrical power and high pressure bottles for gas supply for the ventilator or for oxygen-therapy). Such mobile intensive care stations are equipped with all important devices which are needed for the support of the sick baby, such as ventilators with blender, oxygen monitor, suction device, infusion pump, ECG-Monitor, pulse-oxymeter etc.

Because the environmental conditions may be very severe, sometimes below -10 to -30 °C in Northern countries during the winter, the incubator heat system must be very effective. It is recommended that the small sick baby is kept warm by covering him with a blanket or a silver swaddler. For better fixation during transport the babies can be laid in a vacuum mattress, which can be folded underneath the patient. This will also reduce heat loss.

A transport incubator with double walls or radiant heat source in the wall will substantially reduce an infant's radiant heat loss.

Skin Temperature Control (Servo Control)

An alternative approach to caring for an infant in the neutral thermal environment is to servo the heating device (either the incubator or the open radiant warmer) to the baby's skin temperature. Normally this skin temperature is monitored by a sensor placed on the abdomen if the baby is lying on its back, or over the kidney if lying on its face.

If the skin temperature drops, the warming device increases the power of the heater. This use of servo control can result in very unstable environmental conditions [25] because every change in skin temperature will cause an alteration in the heater power. If, for example, a baby becomes febrile the incubator air temperature will drop, but there is no change in the core temperature. On the other hand, if a baby who is servo controlled dies, its body temperature will be maintained because the incubator temperature will rise !

The temperature probe can easily become detached from the skin and the sensor will then measure the incubator air temperature or that of the open radiant warmer. This can result in inappropriate temperature changes and servo controlled infants should therefore be carefully observed. It is very important that the infant's core temperature is monitored frequently by means of independent thermometers. Modern incubators and open radiant warmers produce an alarm when the skin temperature differs from the set temperature or if there is a malfunction of the skin probe [6, 8].

Influence of the Warming Therapy by other Equipment

If nursed in an incubator, an open radiant warmer or in a heated cot the baby's heat balance can be disturbed by other equipment which is used for the care of the baby but is not intended as warming therapy. This equipment is described in the following sections.

Oxygen Hoods

If the baby needs a higher oxygen concentration in the inspired air, oxygen hoods or head boxes are often used on open radiant warmers or on heated mattresses, when there is no canopy over the mattress. The oxygen has to be humidified and heated because of the heat loss from the head of the baby which is up to 1/4 of the heat loss of the total body!

Oxygen will also dry the respiratory tract. Oxygen tends to fall to the bottom of the hood and flows out without mixing with the air resulting in a very uneven oxygen concentration inside the hood unless it is circulated. The monitored oxygen concentration will depend on the position of the sensor within the hood and may not truly reflect the concentration in the baby's inspired air.

Phototherapy

If phototherapy is used above the baby the short wave length radiant emission goes through the plastic walls of the incubator's canopy resulting in a higher roof temperature and reduced radiant heat loss. The light will heat the skin and can overheat the infant.

The environment inside the incubator will warm up ('greenhouse effect') and the incubator air temperature set value should be reduced by 1.5 to 2 °C before starting the phototherapy to balance this effect. This should also be considered under open radiant warmers.

Breathing Gas Humidifier

If the baby is ventilated, the heat loss by evaporation as a result of humidifying the inspired gas, and the convective heat loss by the temperature difference between the inspired gas and the core temperature cannot be neglected. It is recommended that the inspired gas is warmed and humidified with the temperature monitored near the Y-piece. The humidity of the inspired gas should be between 80 to 100 % r.H (minimum 30 mg H_2O/l according to the ISO Standard [24]). The expired air of a baby has a high humidity of nearly 100 % r.H. and a temperature of 37 °C. If there is a lower ambient humidity the baby will lose water by evaporation. In a the more constant environment of the modern incubator, changes in the breathing gas temperature can be observed to directly influence the core temperature of the baby [12]. The breathing gas temperature should therefore be controlled and monitored. Accumulation of condensed water inside the ventilator tubing can influence the performance of the ventilator and this should be avoided with water traps. Condensation in the ventilator tubing also represents a hygiene risk.

Noise

The canopy of an incubator reduces the noise of the NICU by approximately 10 dB(A) which is a reduction of 50 % in the environmental noise. However the incubator itself can produce its own noise from its fan. The international standard [6] limits this noise to 60 dB(A) but modern incubators have levels clearly lower than this.

There are, however, some other causes of unacceptably high noise levels in an incubator:

- Closing and opening of the port holes can produce a level up to 100 dB(A).
- Putting instruments on the incubator's roof can produce very high peaks.
- The alarm sounds of several medical devices in the NICU can produce an unacceptable noise level and disturb the quiet environment of the baby.
- Tilting mechanism of the mattress tray

Open radiant warmers have no fan noise, but on the other hand there is no canopy which can reduce the noise of the NICU. Some parts of the open radiant warmer can produce unacceptable high noise peaks, for example the closing mechanism of the walls or the mattress tilting. Therefore modern equipment should be used as quietly as possible and nurses should be trained to avoid too much noise.

Conclusion

The better knowledge about the heat losses of the baby in different equipment designed for warming therapy shows us that modern incubators are capable of keeping babies warm, even premature babies of 500 g and at 24 weeks gestation. The importance of the humidity has to be recognized with the need for an active hygienic humidification system and double wall hoods. The disadvantages of the incubator concept with its closed hood are reduced by an improvement in performance when hand ports or front flaps are opened.

Open radiant warmers can also keep premature babies warm. The heating power is sufficient to warm up hypothermic babies in a short time. The main advantage is the free access to the baby, for staff and parents. The disadvantage of higher convective and evaporative heat losses have to be considered if used for long term therapy.

Heated mattresses and cribs are suitable as an alternative method of keeping term babies warm but these methods should not be used for premature babies less than 1500g. Heated mattresses are used in combination with an open radiant warmer for better thermal comfort of the baby. The mattress can only transfer heat if its temperature is different to that of the baby's core temperature.

Several devices used in NICU's disturb the heat balance of the baby, for example oxygen hoods, breathing gas humidifiers and phototherapy. Their influence on the baby has been described and has to be considered during the warming therapy.

Implications for Practice

1. Incubators :

For low birthweight neonates use only modern incubators with:

- low air velocity
- double walls
- high air temperature (39 °C)

Heat shields are not necessary in modern double wall incubators

Hats could be used to avoid unnecessary heat loss

Use modern incubators with an active humidification-system with safe hygienic function (servo-control)

2. Open Radiant Warmers :

- Heat shields should not be used in ORW because they cannot transmit radiant heat to the baby.
- Use the skin control mode for automatic temperature control of the baby

- The manual mode should not be used for more than 15 min

- Skin temperature sensors and their attachment to the baby should be regularly checked.

3. Heated Mattress :

The core temperature follows the mattress temperature:

- Therefore set the mattress temperature to the core temperature or to a higher setting to avoid heat loss.

- Unheated gel or water mattresses in incubators should not be used because of their uncontrolled function (heat loss or gain)

- Heated mattresses should be used in combination with radiant warmers for better comfort of the baby better heat control.

4. Oxygen Hoods :

- Oxygen hoods without circulation should not be used because of the uneven, unknown O_2-concentration inside the hood and because of the effect of dry, cold gases.

5. Phototherapy :

- Phototherapy is a recognizable source of heat for babies.
  This has to be considered and a reduction of the set temperature values (power) of the warming equipment made.

6. Humidifier for Ventilator:

- Use humidifiers with a temperature control for the breathing gas close to the y-piece as this can directly affect the core temperature:

  • heat loss by convection

  • heat loss by evaporation

7. Oxygen-supply:

- Use a remote control for the oxygen supply in incubators to produce safe, constant concentrations.

8. Which type of Equipment should be used ?

			baby's need for			
Equipment	**name**	**baby**	**heat**	**humidity, oxygen**	**care level**	**weight**
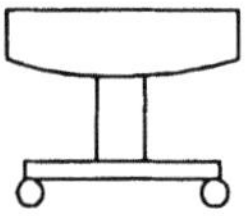	warming bed	not naked	low	--	normal care	> 1500g AGA
	standard incubator sigle wall air temp ≤ 37 °C	naked	medium	yes	normal care	> 1500 g SGA
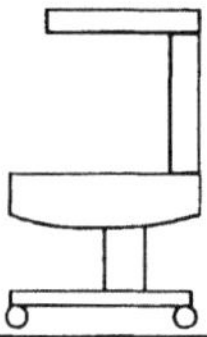	open (intensive) care	naked	high	--	intensive care, observation rescusitation	< 1500 g SGA
	intensive incubator double wall air temp ≤ 39 °C	naked	high	yes	intensive care	< 1500 g SGA

Legend : AGA : Appropriate for gestational age
SGA : Small for gestational age

References:

1. Marx FF (1968) Die Entwicklung der Säuglingsinkubatoren. Verlag Siering KG, Bonn

2. Hammarlund, Sedin, Strömberg Transepidermal Waterloss in Newborn Infants. Acta Paediatrica Scandinavia, Vol. 72, pp 721-728, 1983

3 Brück K Fetal and neonatal physiology. volume 1, chapter 48: Neonatal thermal regulation. figure 48-2, page 490

4. Wasner J. (1995) Heat balance calculation for newborns in incubators. Dräger Medizintechnik aktuell, 1.95, brochure Dräger

5. IEC 601-2-35 Particular requirements for the safety of heated mattresses

6. IEC 601-2-19 Particular requirements for the safety of incubators

7. Evaluation (1984) Infant radiant warmers, Health Devices, Vol 13 No.7

8. IEC 601-2-21 Particular requirements for the safety of radiant warmers

9. Harpin VA, Rutter N (1985) Humidification of incubators. Arch Dis Child 60:219

10. Ducker (1994) Hygiene report with humidification in incubators, submitted to : Archive of Childhood, and brochure Dräger

11. Hey EN, Katz G. (1969) Evaporative water loss in newborn baby. Journal of Physiology, 200

12. Brück (1994) Messungen der Körpertemperaturen von sehr kleinen Frühgeborenen während der ersten Lebenswoche. Dissertation, Universität Düsseldorf

13. Frankenberger H (1991) Inkubatoren. Verlag TÜV Rheinland

14. Sjörs G a.o.(1992) An Evaluation of Environment and Climate Control in Seven Infant Incubators. Biomedical Instrumentation & Technology,

and:

Boone, Otho a.o.: Thermal Evaluation of Four Commercially Available Double Wall Incubators, presented at the 86th Annual Scientific Assembly of the Southern Medical Association

15. Agate FJ, Silverman WA: (1963) The control of body temperature in the small newborn infant by low-energy infra-red radiation. Pediatrics 31:725.

16. Sjörs G, Hammarlund K, Kjartensson S and Sedin G (1994) Thermal balance in term infants nursed in an incubator with a radiative heat source. Pediatric Research 35, (2) 278 abs 115

17. Tunell R (1987) The evaluation of different methods to provide warmth to newborn babies by the use of baby thermal manikin. Zukünftige Entwicklung medizinischer Technik in der Pädiatrie, edited by P. Lemburg and J. Koch, Drägerwerk AG Lübeck

18. Knauth A, Gordin M, NcNelis W, Baumgart S; (1989) Semipermeaable polyurethane membrane as an artificial skin for the premature neonate. Pediatrics 83:945

19. Sarman I, Tunell R (1989) Providing warmth for preterm babies by a heated, water filled mattress. Archives of Disease in Childhood 64: 29-33

and:

Sarman I, Can G, Tunell R (1989) Rewarming preterm infants on a heated, water filled mattress. Archives of Disease in Childhood 64: 687-692

20. Rutter (1992) Temperature control and its disorders. Textbook of neonatology, second edition, Chuchill Livingstone

21. Marshall HK, Avroy AF Care of the high-risk neonate. Chapter 5: The physical environment. Page 114 - 129, W.B. Saunders Company

22. Baumgart S (1986) Partitioning of heat losses and gains in premature newborn infants under radiant warmers. Pediatrics 75: 89

23. Malin S, Baumgart S (1987) Optimal thermal management for low birth weight infants nursed under high powered radiant warmers. Pediatrics 79; 47

24. Standard for safety of humidifiers: ISO 7185

25. Ducker DA, Lyon AJ, Ross RR, Bass CA, McIntosh N (1985) Incubator temperature control: effects on the very low birthweight infant. Archives of Disease in Childhood 60: 902 - 907

26. Okken A, Blijam C, Franz W, Bohn E (1982), Effects of forced convection of heated air on insensible water loss and heat loss in preterm infants in incubators. Journal of Pediatrics 101: 108-112

27. Frankenberger H (1991) Inkubatoren. Verlag TÜV Rheinland, Seite 64 ff

28. Bell EF, et al (1980) Heat balance in premature infants; comparative effects of convectively heated incubator and radiant warmer, with and without plastic heat shield. J. Pediatr 96:460

29. Calder KH, Nardis EE and Ultmann JS (1984) A warming mattress for premature infants. Pediat.Res. 18, 333 A, Abstract 1427

30. Baumgart S, et al (1981) Effect of heat shielding on convective and evaporative heat losses and on radiant heat transfer in premature infant. J Pediatr 99:948

31. Bell EF (1983) Infant incubators and radiant warmers. Early Human Development, 8 , 351-375

32. Gagge AP Standard operative temperature, a generalised temperature scale, applicable to direct and partitional calorimetry, American Journal of Physiology, 31, 1940, p. 93 - 103

33. Topper WH, Stewart TP Thermal support for the very-low-birth-weight infant: Role of supplemental conductive heat: The Journal of Pediatrics, Vol. 105, Number 5, p. 810-814

Chapter 11

Effect of the Thermal Environment on Neonatal Mortality and Morbidity: State of the Evidence

JOHN C. SINCLAIR

Warmth, as well as food and adequate oxygen, is a fundamental need of the newborn infant. Although this may seem self-evident, it is only within the last 35 years that evidence concerning the precise thermal requirements of babies has been obtained.

In 1975, I reviewed evidence from observational and experimental studies concerning the effect of the thermal environment on neonatal mortality and morbidity [1]. This was followed in 1992 by a more systematic, quantitative review of evidence from randomized trials [2] using meta-analysis. These reviews indicate that the thermal environment has important effects not just on body temperature and physiologic responses of human neonates, but also on clinical outcomes including neonatal death rate.

However, questions are being raised as to whether this body of evidence is applicable to the very immature, very small babies who are presently surviving in vastly increased numbers. Such infants were barely represented in previous randomized trials which assessed the effect of the thermal environment on clinically important outcomes. This caution should be kept in mind as the available evidence is reviewed in this report.

The objectives of the present report are to summarize currently available evidence from randomized trials concerning the effect of the thermal environment on neonatal mortality and morbidity, to indicate the limits of applicability of such evidence, and to identify research objectives for the future.

Neonatal Heat Loss

Before birth, the temperature of the human fetus is about 0.5 °C higher than the maternal body temperature - a gradient that allows the dissipation of the heat of fetal metabolism. After birth, thermal interchanges with the environment occur through radiation, convection, conduction and evaporation. When the rate of heat loss by these routes exceeds the rate of heat production, the baby's temperature falls.

Newborn infants are prone to excessive heat loss because, being small, they have poor thermal insulation. Moreover, they have a large surface area in relation to body mass. These disadvantages are particularly marked in the low birthweight baby. The addition of clothing and blankets restores total insulation of the full-size baby to that typical of the clothed adult. Thus, the clinical problem of managing the thermal environment arises in babies of any size when they are exposed, naked, to the environment in order to permit careful observation during intensive care, and in babies of low birthweight. As will be discussed in this review, important negative effects - above all an increase in death rate - have been shown to result when appropriate measures are not taken to limit the magnitude of heat loss of babies during the first days after birth.

For babies nursed in single-walled incubators, the major route of heat loss is radiation, whereas under a radiant warmer, radiation becomes an avenue of heat gain. The proportion of metabolic heat production which is lost by evaporation, about 30 % in the incubator, rises to about 75 % under a radiant warmer. High evaporative water losses can lead to hypertonic dehydration unless such increased water losses are quantitatively replaced. Whether the replacement by intravenous infusion of high evaporative water losses leads to an increased risk of symptomatic patent ductus arteriosus (PDA), and whether lessening the problem by providing a high-humidity environment would reduce the incidence of symptomatic PDA is an attractive hypothesis but it has not been tested experimentally. Other clinical effects of the specific route of heat loss, as distinct from the total magnitude of heat loss, have not been identified.

In a thermal environment providing no gradient for heat loss by radiation or convection, heat must be lost mainly by evaporation. However, sweating capacity is limited or even absent in pre-term infants. Thus, serious hyperthermia may result if a pre-term infant is placed in an environment which is too warm.

Neonatal Thermoregulation

The resting rate of heat production of the newborn baby increases from a typical range of 35-40 kcal/kg/day on day one to 40-60 kcal/kg/day during the later neonatal period. Healthy newborn infants increase their rate of heat production in response to cold stimulation, mainly by non-shivering thermogenesis in brown adipose tissue. Non-shivering thermogenesis is triggered when the mean skin temperature is lowered to 35-36 °C.

A "thermoneutral" condition is one which incites no thermally-induced metabolic response from the baby, body temperature being regulated only by the physical processes of vasoconstriction and/or vasodilatation. There have been several approaches to defining the thermoneutral state for newborn infants, based on environmental temperature, skin-environmental temperature gradient, or anterior abdominal skin temperature. Of these, the last is the only approach which has been tested experimentally. For babies <1500g nursed in incubators, a neutral thermal state is achieved when incubator heating is adjusted to maintain the abdominal skin temperature between 36 and 37 °C [3]. For babies under radiant warmers, it appears likely that a thermoneutral condition is achieved at an abdominal skin temperature between 36.5 and 37.5 °C [4]; however, further research is needed.

The environmental temperature or skin-environmental temperature gradient required for thermoneutrality is not a constant, but varies with the baby's size and age. Charts have been constructed which predict the thermoneutral environmental temperature range for babies of differing weight and age. Typical values for the "operative" environmental temperature required to provide reasonable warmth for naked low birthweight babies range from 31-36 °C depending on the baby's weight and postnatal age. Even warmer temperatures may be required for babies born at the current threshold of viability.

Thermal Protection of Newborn Infants

Thermal Protection at Birth

Cold exposure at birth has a number of effects on the newborn infant [1]. It plays at least a contributory role in the establishment of air breathing. Cold exposure produces a rise in metabolic rate, a slower recovery from metabolic acidosis following labor and delivery, and a lower arterial oxygen tension during the first hour after birth compared to babies subjected to less cold exposure. The early postnatal rise in glycerol and free fatty acid concentrations, indicators of lipolysis, is more pronounced with cold stimulation. Cooling appears to play a contributory role in the early postnatal rise in TSH concentration. The effect of body temperature on drug metabolism and drug half-life in human newborns has been little studied.

The most-studied effect of cold exposure at birth is the effect on heat loss and body temperature per se. Immediately following birth, there are large evaporative and non-evaporative heat losses from the baby's warm, wet skin. Unless precautions are taken to limit these losses, the baby's core temperature, which is about 37.8 °C at birth, may fall by 2-3 °C or more during the first 30-60 minutes.

A fall in the baby's rectal temperature exceeding 1-2 °C should be avoided. Simple measures to reduce heat loss at birth include drying the baby and either swaddling in warmed blankets or skin-to-skin contact between mother and baby under warmed blankets [5-17]. These measures are sufficient for vigorous, full-size babies and do not interfere with early mother-infant contact or with the early opportunity for suckling. For babies who require immediate interventions at birth (e.g. resuscitation), such procedures should be performed under a radiant heater. No elective procedures which may entail cold stress (e.g. bathing, circumcision) should be undertaken immediately after birth.

Thermal Protection in the Newborn Nursery

During the first days and weeks of life, there are a number of effects of the thermal environment on the neonate which continue to be important [1]. The best studied and most important of these is the effect of moderate differences in the thermal environment on the survival rate of low birth-weight infants (see below). A specific clinical syndrome, neonatal cold injury, is the most dramatic expression of the lethal effect of prolonged cold exposure. This clinical entity was first clearly described in the U.K. in the 1950's and remains an important problem in developing countries today. Babies who die of neonatal cold injury usually have massive pulmonary hemorrhage, and infection is also common. Hypothermic babies commonly have defects of blood coagulation and an increased risk of clinical bleeding.

For full-size babies, the provision of clothing (shirt, diaper and gown) and bedding (flannelette sheet and cotton blanket) allows the baby to achieve thermal balance with minimal rate of heat production at an air temperature of 24 °C. Clothed, low birthweight babies who are nursed in cots will require higher air temperatures (in the case of <1500g babies, up to 30 °C) if they are to maintain normal body temperature with minimal rate of heat production. A heated water-filled mattress, whose temperature can be regulated between 35-38 °C, is a useful adjunct to the care of low birthweight infants in cots [18, 19].

For low birthweight babies in incubators, controlled trials have been carried out which test the effect of varying relative humidity, air temperature or abdominal skin temperature on clinical outcomes including mortality. It has been shown that increasing the relative humidity in cool incubators (mean incubator air temperature 29 °C) resulted in an important reduction in mortality among low birthweight infants [20], table 11.-1 (This trial included some babies under 1000g birthweight, but the numbers of participants in this weight group, and their results, were not separately reported.). Because body temperatures were higher in the high-humidity group, it was hypothesized that the high-humidity environment caused a reduction in (evaporative) heat loss, and that the effect on mortality was actually a thermal effect. This hypothesis was supported by the results of a subsequent trial in which the effects of the two humidity conditions were compared in groups of infants maintained at the same body temperature; under these conditions, there was no difference in mortality [21]. (In this trial, 96/362 (27 %) were ≤1000g birth weight but the results were not separately reported for this subgroup.) Thus, there is no survival advantage attached to high humidity per se provided that appropriate thermal conditions are provided by other means.

Tab. 11.-1

Effect of Increasing the Relative Humidity (RH) in Cool Incubators

	RH 80-90% n (%)	RH 30-60% n (%)	ERR (95% CI)	ERD (95% CI)
Death (trial entry to 5 days)				
Silverman [20]	36/164 (22.0)	56/160 (35.0)	0.63 (0.44,0.90)	-13.0 (-22.8,-3.2)

ERR = event rate ratio
ERD = event rate difference
CI = confidence interval

Two randomized trials of different air temperatures within incubators demonstrated a survival advantage of the warmer air-temperature condition [22, 23], table 11.-2. The air temperatures studied in these trials would be considered too cool by today's standards - even the warmer condition in one trial was only 31.1 - 32.2 °C. The overview of these two trials shows that the warmer air-temperature condition caused a substantial and highly significant reduction in death rate, computed from trial entry to age 21-28 days: event rate ratio 0.60, 95 per cent CI 0.45, 0.82; event rate difference -15.8 per cent, 95 per cent CI -24.8, -6.7. Thus, the warmer condition prevented 1 death for each 6 infants treated (95 per cent CI 4, 15).

In the American trial by Silverman et al [22] reviewed above, 28 of 91 participants (31 %) weighed less than 1000g at birth. The relative risk reduction for death to five days in this subgroup, 42 %, was very similar to the relative risk reduction for this outcome among all trial participants, 48 %. On the other hand, only 13 of 214 participants (6 %) in the British trial of Jolly et al [23] weighed under 1000g at birth, and none of these 13 survived to 21 days in either treatment group.

Tab. 11.-2

Effect of Increasing the Air Temperature of Cool Incubators

	Higher air temperature[a] n (%)	Lower air temperature[b] n (%)	ERR (95% CI)	ERD % (95% CI)
Death (trial entry to 21-28 days)				
Silverman[c] [22]	20/91 (22.0)	41/91 (45.1)	0.49 (0.31,0.76)	-23.1 (-36.4,-9.8)
Jolly[d] [23]	28/109 (25.7)	37/105 (35.2)	0.73 (0.48,1.10)	-9.6 (-21.9,2.7)
Typical estimates			0.60 (0.45,0.82)	-15.8 (-24.8,-6.7)

a Higher air temperature: Silverman, 31.7 °C (range 31.1 - 32.2 °C): Jolly, manually adjusted to keep rectal temperature at 35.6 - 37.2 °C

b Lower air temperature: Silverman 28.9 °C (range 28.3 - 29.4 °C), Jolly, 29.4 °C

c Death before 28 days

d Death before 21 days

The "warm" condition tested in the above trials was still below the thermoneutral state. Two subsequent controlled trials tested the effect on the death rate of low birthweight infants of maintaining abdominal skin temperature at 36 °C by servo-controlled radiant heat in addition to "warm" incubator air temperature [24, 25], table 11.-3. The overview of the two trials shows that servo-controlling abdominal skin temperature at 36 °C produced a further reduction in death rate. This effect was particularly marked in <1500g babies: event rate ratio 0.66, 95 per cent CI 0.48, 0.90; event rate difference -18.7 per cent, 95 per cent CI -31.2, -6.2. Thus, the servo-controlled condition prevented 1 death for each 5 infants treated (95 per cent CI 3, 16).

Tab. 11.-3 : Effect of Maintaining Abdominal Skin Temperature at 36 °C Using Servo-controlled Radiant Heat

Effect of Maintaining Abdominal Skin Temperature at 36 °C Using Servo-controlled Radiant Heat

	Abdominal skin servo-controlled at 36°C n	(%)	No servo-control of abdominal skin temperature[a] n	(%)	ERR (95% CI)	ERD (95% CI) %
Death (very low birth weight)[b, c]						
Day [24]	6/33	(18.2)	18/37	(48.6)	0.37 0.17, 0.83	-30.5 (-51.3,-9.7)
Buetow [25]	37/89	(41.6)	37/69	(53.6)	0.78 0.56, 1.08	-12.1 (-27.8, 3.6)
Typical estimates					0.66 0.48, 0.90	-18.7 -31.2, -6.2

a In each thermal condition, incubator air temperature set at 31.8 °C (range 31.1 - 32.2 °C)

b Day, death before discharge from hospital; Buetow, death before 28 days

c Day, 800-1599g; Buetow, 1000-1500g

In the trial of Day et al [24] reviewed above, 70 of 125 participants (56 %) weighed 800-1599g at birth. These babies accounted entirely for the reduction in mortality found in this trial. Servo-control of body temperature caused no reduction in mortality in larger babies.

The results of this series of trials suggest that, in the aggregate, the careful control of the thermal environment in incubators has the capacity to reduce the neonatal mortality rate of very low birthweight infants by at least 25 per cent (absolute risk reduction) below the rate experienced in the mid-1950's. This is probably the largest effect on mortality ever demonstrated in the field of neonatology. By comparison, for example, meta-analysis of surfactant trials shows that prophylactic surfactant causes an absolute reduction in death rate of the order of 5-6 % [26].

With the exception of the trial of Jolly et al [23] in which birthweight under 1000g was uniformly fatal, there is no suggestion in these trials that the benefits of thermal protection did not extend to the smallest babies. In fact, if anything, the results suggest that it is in the smallest babies that the survival advantage due to thermal protection was largest in absolute terms.

However, it must be recalled that the mechanism of the protective effect of the warmer thermal environment was not uncovered in these trials. No distinctive or unique cause of death was selectively reduced by thermal protection. One might speculate that the energy cost of defending deep body temperature against cold contributes to the lethal effect, particularly in the first days of life, when the balance between caloric supply and caloric expenditure is tenuous, and particularly in the smallest infants whose glycogen and fat stores are most limited. However, it is also possible that lower body temperature per se is the problem, perhaps by slowing the rate of critical enzymatic reactions. For example, the coagulation cascade is strongly inhibited by hypothermia [7].

There are major differences in non-thermal aspects of neonatal care between the 1950's and 1960's, when the above-cited trials were performed, and current practice. For example, babies are no longer starved and thirsted for 48-72 hours after birth. Whether an effect of the thermal environment on survival could be shown if the trials were to be repeated today is unknown, and this uncertainty will persist because the mechanism of the protective effect of warmth has not been proved. Nevertheless, it is prudent to assume that regulation of the thermal environment to achieve thermoneutral conditions continues to be a major requirement for effective care of the low birthweight or sick neonate.

Radiant warmers were introduced into newborn nurseries primarily because they offered easier access to the baby receiving intensive care than did the enclosed incubator. No controlled evaluation of the effects of radiant warmers vs incubators on neonatal mortality or morbidity has been reported. However, several controlled trials have compared radiant warmers with incubators with respect to their effects on intermediate outcomes including aspects of thermal balance and bacterial colonization. Radiant warmers, compared to incubators, increase the insensible water loss of low birthweight infants by 0.94 g/kg/h, 95 per cent CI 0.48, 1.40 [28-30], table 11.-4. This leads to an increased water requirement amounting, on average, to 23 ml/kg/day (95 per cent CI 12, 34) which should be taken into account in prescribing the daily water intake for babies nursed under radiant warmers. Radiant warmers, compared to incubators, produce a slight increase in the rate of the baby's oxygen consumption, but the difference does not appear to be significant either statistically or clinically: mean increase 0.27 ml/kg/min, 95 per cent CI -0.09, 0.63 [29-33], table 11.-5. In the one controlled trial which assessed bacterial colonization, there was no difference between radiant warmers and incubators in the colonization rate or in the distribution of colonizing organisms [34].

The high evaporative water losses under radiant warmers can be reduced by the use of a loosely fitting Saran plastic blanket placed over the baby [35, 36]. This reduces evaporative water loss without interfering with the delivery of radiant heat to the baby. For babies in incubators, evaporative water loss can be reduced by raising the ambient humidity [37, 38], or by placing a plastic heat shield [29] or thermal blanket [39] over the baby.

Tab. 11.-4 : Effect of Radiant Warmer vs Incubator on Rate of Insensible Water Loss (g/kg/h)

	Radiant Warmer			Incubator				
	n	x	s.d.	n	x	s.d.	MD	95 % CI
Bell [28]	10	2.43	0.76	10	1.58	0.82	0.85	0.16, 1.54
Bell [29]	7	3.40	1.50	8	2.37	1.15	1.03	-0.34, 2.40
Marks [30]	9	1.98	0.75	9	0.97	0.78	1.01	0.30, 1.72
Weighted mean difference							0.94	0.48, 1.40

MD = mean difference

Tab. 11.-5 : Effect of Radiant Warmer vs Incubator on Rate of Oxygen Consumption (ml/kg/min)

	Radiant Warmer			Incubator				
	n	x	s.d.	n	x	s.d.	MD	95 % CI
Darnall [31]	10	6.33	1.80	10	6.23	1.69	0.10	-1.43, 1.63
Bell [29]	7	7.73	0.52	8	7.38	0.93	0.35	-0.39, 1.09
Marks [30]	9	5.77	0.50	9	5.54	0.58	0.23	-0.28, 0.74
LeBlanc [32]	16	7.45	1.76	16	6.84	1.48	0.61	-0.51, 1.73
Marks [33]	13	8.64	1.30	13	8.67	1.80	-0.03	-1.24, 1.18
Weighted mean difference							0.27	-0.09, 0.63

The importance of the provision of a thermoneutral environment is greatest during the first week of life. After the first week, there is less evidence that the thermal environment influences mortality. Indeed, young infants inevitably will face cold stress. There is evidence from two controlled trials [40, 41] concerning the effects of exposure of low birthweight infants to a slightly sub-neutral environment between 1-3 weeks postnatal age. Compared to infants whose abdominal skin temperature was servo-controlled at 36.5 °C, servo-control at 35 °C results in enhanced capacity to defend deep body temperature during a subsequent cold stress, but a slower rate of growth. By increasing the energy intake by the amount needed to offset the energy cost of the cooler condition (+8 kcal/kg/day), the growth deficit can be repaired. Thus, low birthweight infants raised under slightly sub-thermoneutral conditions during the late neonatal period and provided with a slightly increased milk intake have the advantages of both a rapid weight gain and increased ability to resist an acute cold stress.

Thermal Protection at Home

Home care of low birthweight infants has been undertaken, particularly in the developing world, utilizing the body heat of the mother as a heat source for the baby. The method consists of placing the baby, naked except for diapers, between the mother's breasts. The baby is carried in a vertical position next to the mother's skin in a marsupial-like pocket formed by the mother's loose-fitting blouse or dress. The method was introduced in an effort to allow early discharge of low birthweight babies from an overcrowded hospital and to promote bonding and lactation. The evidence available from controlled trials of so-called "kangaroo baby care" is very limited at present [42-44]. Large scale randomized trials, particularly in the developing world, are needed in order to provide a valid basis for assessing the thermal and non-thermal effects of this practice. Any recommendation concerning the efficacy and safety of kangaroo baby care must await the completion of such trials.

Treatment of Hypothermia

Babies who develop severe hypothermia should be rapidly rewarmed so that they attain normal body temperatures within a few hours. This is facilitated by using a thermostatically-controlled heated mattress set at 37-38 °C in addition to measures to reduce heat losses [45, 46].

Future Research

The limited applicability of the conclusions of past research to the present populations of surviving extremely immature infants has been appropriately questioned by some. However, we should not "throw out the baby with the bath water". First, it is highly plausible that the current successes in increasing the survival rate of very small babies are being achieved as a result of (not in spite of) applying the conclusions of research conducted mainly in somewhat larger infants. Second, we should note a major strength of previous research concerning management of the thermal environment of the newborn: the research question has extended to asking whether the intervention matters in terms of clinically important outcomes, not merely physiologic responses. Thus, in framing new research questions in new populations of infants, we should require a similar standard. It will not be sufficient simply to demonstrate the technical feasibility of innovations in thermal management or to observe thermoregulatory responses of the baby to such innovations. For example, it is now possible to measure and compare body temperatures taken simultaneously at two or more sites, and to continuously record temperatures with trend display. Whether these innovations truly constitute an advance will depend on the answer to the question of whether babies are better off in terms of their life and health when their care is monitored in such ways. Thus, randomized trials of the "utility" of such innovations will be required in future.

Summary

Healthy newborn infants, including low birthweight infants, are capable of making the appropriate physiologic responses for the regulation of their body temperature. However, because of their small size, newborn babies are inefficient at limiting heat losses when exposed to cold stress. Thus, during the first days and weeks following birth, the prevention of excessive thermal stress is of critical importance in newborn care. Evidence from randomized trials shows that the careful regulation of the thermal environment to provide appropriately warm conditions reduces the death rate of low birthweight infants. Although extremely premature infants (e.g. <800g birth weight, <26 weeks gestation) were barely represented in trials testing the effect of the thermal environment on survival, there is little basis in rationale or evidence to suggest that this conclusion does not apply also to such babies.

Implication for Practice

1. Heat loss at birth. A fall in the baby's rectal temperature exceeding 1-2 °C should be avoided. Simple measures to reduce heat loss from the newborn include drying and either swaddling in warmed blankets or skin-to-skin contact between mother and baby under warmed blankets. These measures are sufficient for vigorous, full-sized babies.

2. Thermal protection for low birth-weight babies. During at least the first week after birth, low birthweight babies should be provided with a carefully regulated thermoneutral environment. For naked babies in incubators, this is achieved by adjusting incubator heating to maintain abdominal skin temperature at approximately 36.5 °C, using either servo-control or frequent manual adjustment. For naked babies under radiant warmers, this is achieved by adjustment of the radiant heater output to maintain the abdominal skin temperature at about 37 °C. For clothed, blanket-covered babies in cots, this is achieved by adjusting the room temperature between 25-30 °C (depending on the baby's weight and age), or by using a heated mattress whose temperature is adjustable within the range of 35-38 °C; rectal temperature should be maintained at about 37 °C.

3. Reduction of high evaporative water and heat losses. Evaporative water and heat losses are increased under radiant warmers and much increased in naked, very preterm babies nursed either in incubators or radiant warmers. Under radiant warmers, the use of a plastic blanket which is loosely fitted over the baby reduces evaporative water loss without interfering with the delivery or radiant heat to the baby. For babies in incubators, evaporative water loss can be reduced by raising the ambient humidity, or by placing a plastic heat-shield or thermal blanket over the baby.

4. Warming hypothermic babies. Rapid warming of hypothermic babies should be undertaken. This is facilitated by using a thermostatically-controlled heated mattress set at 37-38 °C in addition to measures to reduce heat losses.

Recommendations for Future Research

1. Future trials of alternative methods of providing immediate thermal protection at birth should be carried out in low birthweight infants.

2. More evidence from controlled trials is needed to define the thermoneutral state for babies nursed under radiant warmers.

3. The thermal and non-thermal advantages and disadvantages of radiant warmers vs incubators should be addressed in future trials.

4. The effects of high vs low humidity in incubators on the incidence of symptomatic patent ductus arteriosus should be assessed in future trials carried out under conditions which provide equivalent heat balance and water balance.

5. Innovations in incubator design should be tested in randomized trials for their effects on clinically important morbidity and on mortality.

6. The physics of heat transfer during skin-to-skin contact as practised in kangaroo baby care should be described quantitatively over a range of newborn weights and environmental temperatures. Large-scale randomized trials, particularly in the developing world, should address the thermal and non-thermal effects of kangaroo baby care of low birthweight infants.

References

1. Sinclair JC (1975) The effect of the thermal environment on neonatal mortality and morbidity. In: Preventability of Perinatal Injury, Ed. K. Adamsons and H.A. Fox, Alan R., Liss, Inc. New York

2. Sinclair JC (1992) Management of the thermal environment. In Effective Care of the Newborn Infant. eds. JC Sinclair, MB Bracken, Oxford University Press, Oxford, Chapter 4, pp 40-58.

3. Silverman WA, Sinclair JC, Agate FJ. (1966) The oxygen cost of minor changes in heat balance of small newborn infants. Acta Paediatr Scand ; 55:294-300.

4. Baumgart S (1985) Partitioning of heat losses and gains in premature newborn infants under radiant warmers. Pediatrics ; 75:89-99.

5. Gandy GM, Adamsons K Jr, Cunningham N, Silverman WA, James LS (1964) Thermal environment and acid-base homeostasis in human infants during the first few hours of life. J Clin Invest ; 43:751-758.

6. Miller DL, Oliver TK (1966) Body temperature in the immediate neonatal period: The effect of reducing thermal losses. Am J Obstet Gynecol ; 94:964-969.

7. Baum JD, Scopes JW (1968) The silver swaddler; device for preventing hypothermia in the newborn. Lancet ; 1:672-673.

8. Stephenson JM, Du JN, Oliver TK (1970) The effect of cooling on blood gas tensions in newborn infants. J Pediatr ; 76:848-852.

9. Besch NJ, Perlstein PH, Edwards NK, Keenan WJ, Sutherland JM (1971) The transparent baby bag. A shield against heat loss. New Engl J Med ; 284:121-124.

10. Dahm LS, James LS (1972) Newborn temperature and calculated heat loss in the delivery room. Pediatrics ; 49:504-513.

11. Hobbs JF, MacKuanying N, Eidelman AI, Schneider KM (1975) A new transparent insulating infant gown to maintain thermal stability. Pediatr Res ; 9:366.

12. Omene JA, Diejomaoh FME, Faal M, Diakparomre MA, Obiaya M (1979) Heat loss in Nigerian newborn infants in the delivery room. Int J Gynaecol Obstet ; 16:300-302.

13. Chaput De Saintonge DM, Cross KW, Hathorn MK, Newis SR, Stothers JK (1979) Hats for the newborn infant. Br Med J ; 2:570-571.

14. Fardig JA (1980) A comparison of skin-to-skin contact and radiant heaters in promoting neonatal thermoregulation. J Nurse-Midwifery ; 25:19-28.

15. Holzman IR (1985) A method to maintain infant temperature. Am J Dis Child ; 139:3990-392.

16. Greer PS (1988) Head coverings for newborns under radiant warmers. J Obstet Gynecol Neo Nurs ; 4:265-271.

17. Johanson RB, Spencer SA, Rolfe P, Jones P, Malla DS (1992) Effect of post-delivery care on neonatal body temperature. Acta Paediatr ; 81:859-863.

18. Sarman I, Tunell R (1989) Providing warmth for preterm babies by a heated, water filled mattress. Arch Dis Child ; 64:29-33.

19. Sarman I, Tunell R (1989) Rewarming preterm babies by a heated, water filled mattress. Arch Dis Child ; 64:687-92.

20. Silverman WA, Blanc WA (1957) The effect of humidity on survival of newly born premature infants. Pediatrics ; 20:477-87.

21. Silverman WA, Agate FJ, Fertig JW (1963) A sequential trial of the nonthermal effect of atmospheric humidity on survival of newborn infants of low birth weight. Pediatrics ; 31:719-724.

22. Silverman WA, Fertig JW, Berger AP (1958) The influence of the thermal environment upon the survival of newly born premature infants. Pediatrics ; 22:876-886.

23. Jolly H, Molyneux P, Newell DJ (1962) A controlled study of the effect of temperature on premature babies. J Pediatr ; 60:889-894.

24. Day RL, Caliguiri L, Kamenski C, Ehrlich F (1964) Body temperature and survival of premature infants. Pediatrics ; 34:171-181.

25. Buetow KC, Klein SW (1964) Effect of maintenance of "normal" skin temperature on survival of infants of low birth weight. Pediatrics ; 34:163-170.

26. Soll RF, McQueen MC (1992) Respiratory distress syndrome. In Effective Care of the Newborn Infant, ed. JC Sinclair, MB Bracken, Oxford University Press, Oxford, Chapter 15, pp 325-358.

27. Rohrer MJ, Natale AM (1992) Effect of hypothermia on the coagulation cascade. Crit Care Med ; 20:1402-1405.

28. Bell EF, Neidich GA, Cashore WJ, Oh W (1979) Combined effect of radiant warmer and phototherapy on insensible water loss in low-birth-weight infants. J Pediatr ; 94:810-813.

29. Bell EF, Weinstein MR, Oh W (1980) Heat balance in premature infants: Comparative effects of convectively heated incubator and radiant warmer, with and without plastic heat shield. J Pediatr ; 96:460-465.

30. Marks KH, Gunther RC, Rossi JA, Maisels MJ (1980) Oxygen consumption and insensible water loss in premature infants under radiant heaters. Pediatrics ; 66:228-232.

31. Darnall RA, Ariagno RL (1978) Minimal oxygen consumption in infants cared for under overhead radiant warmers compared with conventional incubators. J Pediatr ; 93:283-287.

32. LeBlanc MH (1982) Relative efficacy of an incubator and an open warmer in producing thermoneutrality for the small premature infant. Pediatrics; 69:439-445.

33. Marks KH, Nardis EE, Momin MN (1986) Energy metabolism and substrate utilization in low birth weight neonates under radiant warmers. Pediatrics ; 78:465-472.

34. Merenstein GB, Koziol DF, Brown GL, Weisman LE (1979) Radiant warmers vs incubators for neonatal care. Am J Dis Child ; 133:857-858.

35. Baumgart S, Fox WW, Polin RA (1982) Physiologic implications of two different heat shields for infants under radiant warmers. J Pediatr ; 100:787-790.

36. Baumgart S, Engle WD, Fox WW, Polin RA (1981) Effect of heat shielding on convective and evaporative heat losses and on radiant heat transfer in the premature infant. J Pediatr ; 99:948-956.

37. Hey EN, Katz G (1969) Evaporative water loss in the newborn baby. J Physiol ; 200:605-619.

38. Sauer PJJ, Dane JH, Visser HKA (19984) Influence of variations in the ambient humidity on insensible water loss and thermoneutral environment of low birth weight infants. Acta Paediatr Scand ; 73:615-619.

39. Darnall RA, Ariagno RL (1979) Resting oxygen consumption of premature infants covered with a plastic thermal blanket. Pediatrics ; 63:547-551.

40. Glass L, Silverman WA, Sinclair JC (1968) Effect of the thermal environment on cold resistance and growth of small infants after the first week of life. Pediatrics ; 41:1034-1046.

41. Glass L, Silverman WA, Sinclair JC (1969) Relationship of thermal environment and caloric intake to growth and resting metabolism in the late neonatal period. Biol Neonat ; 14:324-340.

42. Whitelaw A, Heisterkamp G, Sleath K, Acolet D, Richards M (1988) Skin to skin contact for very low birthweight infants and their mothers. Arch Dis Child ; 63:1377-1381.

43. Christensson K, Siles C, Moreno L, Belaustequi A, De La Fuente P, Lagercrantz H, Puyol P, Winberg J (1992) Temperature, metabolic adaptation and crying in healthy full-term newborns cared for skin-to-skin or in a cot. Acta Paediatr ; 81:488-493.

44. Christensson K (1994) Fathers are as effective as incubators for babies born by Caesarean section. Appendix IV in Christensson K. Care of the Newborn Infant. Satisfying the need for comfort and energy conservation, doctoral thesis, Karolinska Institute, Stockholm.

45. Sarman I, Can G, Tunnell R (1989) Rewarming preterm infants on heated, water filled mattress. Arch Dis Child ; 64:687-692.

46. Tafari N, Gentz J (1974) Aspects on rewarming newborn infants with severe accidental hypothermia. Acta Paediatr Scand ; 63:595-600.

Chapter 12

Treatment of Sick Newborns in Incubators

EDWARD F. BELL

Introduction

Incubators were first used for the care of preterm infants in the middle of the nineteenth century [1, 2]. In the ensuring years, much has been learned about the safe and effective use of incubators to provide the thermal support needed by preterm and sick infants [3, 4]. In the past 25 years, open radiant warmers have also been found to provide a safe, effective thermal environment for such infants.

Incubator vs Radiant Warmer

When compared with radiant warmers, incubators offer several advantages that may be of physiological importance to preterm and sick infants. First, although incubators provide a steady drone of background noise at a level of 50 to 55 dB(A) [3], primarily from the fan motor, the incubator hood insulates the infant from intermittent loud noises in the nursery environment. Second, the incubator allows lower insensible water loss than from the same infant under a radiant warmer [5-7]. This increase in insensible water loss with the radiant warmer is most likely explained entirely by the observation that ambient water vapor pressure is lower under the radiant warmer than in an incubator [7, 8].

On the other hand, incubators have two important disadvantages compared with radiant warmers. First, some of the smallest, youngest preterm infants cannot be adequately warmed in an incubator because the required air temperature is higher than the incubator's controlling software allows [9]. Second, whenever the incubator's door is opened to allow access to the infant, the environmental temperature drops precipitously [3, 4].

The first of these disadvantages is illustrated in figure 12.-1, which shows the incubator and body tempereatures and body heat balance of two hypothetical infants in incubators, one weighing 1.5 kg and the other 0.5 kg. The heat gain and heat loss of the 1.5 kg infant (Figure 12.-1, left) are equal at an incubator air temperature of 34 °C. In contrast, the heat loss exceeds heat gain for the 0.5 kg infant (Figure 12.-1, right), even with the incubator air temperature at ist maximum of 38 °C. The inadequate thermal support provided by incubators to such very small, young preterm infants, can be overcome by reducing heat loss (or allowing net gain of heat) by any of the four routes of heat exchange with the environment: convection, evaporation, radiation or conduction.

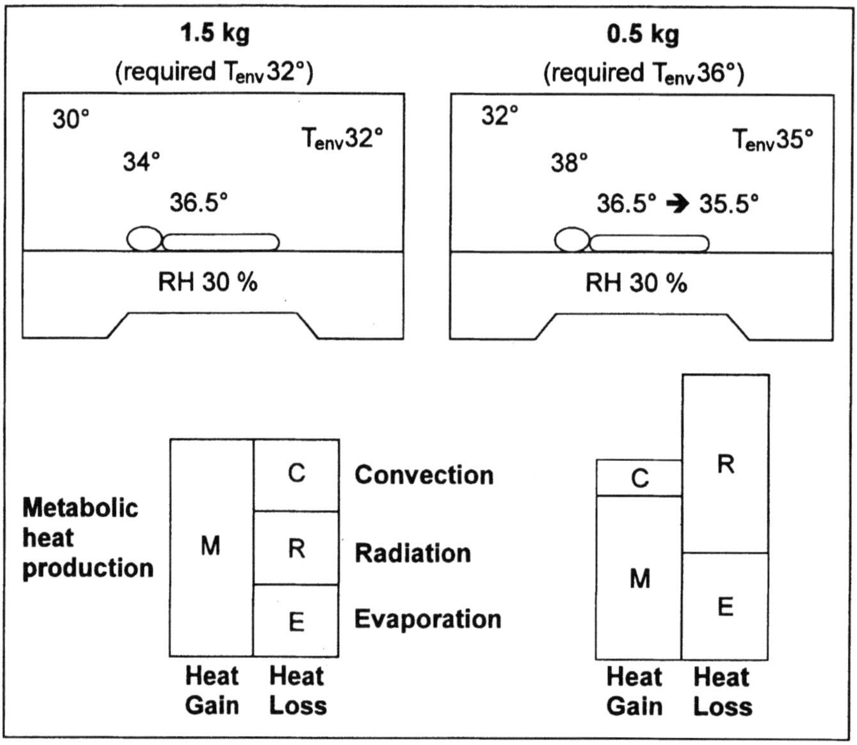

Fig. 12.-1 : The upper half shows environmental and body temperatures of two hypothetical preterm infants in a single-walled incubator, a 1.5 kg infant (left) and a 0.5 kg infant (right). The incubator temperatures shown are inner wall, air, and T_{env}, ,which indicates operative environmental temperature - a weighted mean of inner wall and air temperatures. Also shown is the trunk surface temperature of the infant. RH indicates relative humidity. The lower half shows the components of heat gain and heat loss from the infants' bodies, expressed in watts per kg. M is the infant's intrinsic metabolic heat production. C is heat loss (or gain) by convection, which is directly proportional to the difference between skin and air temperatures. R is heat loss (or gain) by radiation, which is directly proportional to the difference between skin and inner wall temperature. E is evaporated heat loss, which is inversely proportional to body weight and to absolute humidity, which in turn depends on relative humidity and air temperature. The 1.5 kg infant is in thermal balance at an air temperature of 34 °C, but the 0.5 kg infant is in thermal balance even at the highest possible air temperature of 38 °C.

Convective heat losses can be eliminated by increasing the incubator air temperature to the highest possible level, which is typically 38 °C although sometimes as low as 37 °C [10]. In order to achieve a high enough operative environmental temperature for a very small infant, it may be necessary to adjust the air temperature to a level above the incubator's normal maximum by overriding the incubator's high temperature limit (Figure 12.-2). *Evaporative* losses can be reduced by increasing the ambient humidity within the incubator (Figure 12.-3) [11, 12]. *Radiant* losses can be reduced by use of a hood warmer [13] on the outside of the incubator hood or a heat shield inside the incubator (Figure 12.-4) [14]. *Conductive* heat loss is ordinarily considered to be negligible in infants in incubators, but the infant's own metabolic heat can be augmented by the gain of conductive heat from a heated mattress (Figure 12.-5) [15, 16].

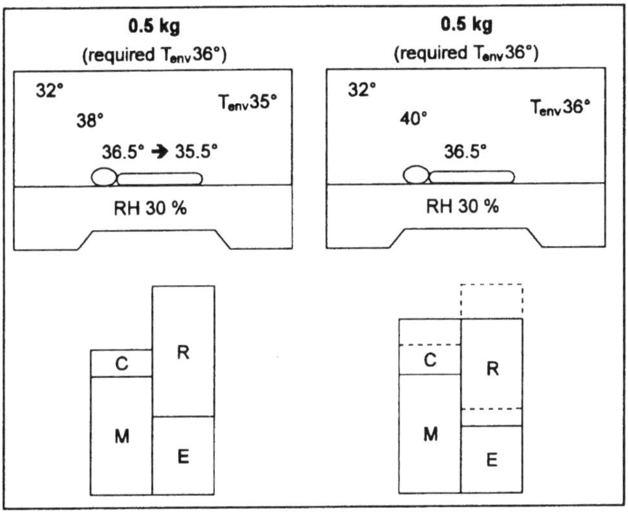

Fig. 12.-2 : This figure illustrates correction of the heat imbalance of the 0.5 kg infant by overriding the maximum air temperature and increasing it further to 40 °C, thereby increasing heat gain by convection and reducing heat loss by radiation.

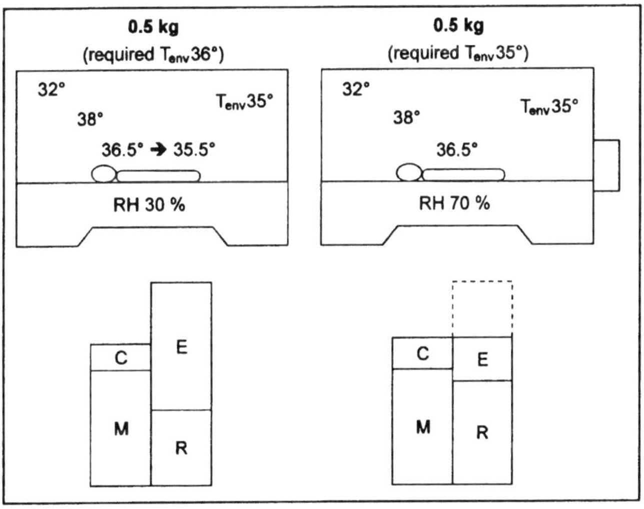

Fig. 12.-3 : This figure illustrates correction of the heat imbalance of the 0.5 kg infant by adding a humidifier that increases the relative humididy from 30 % to 70 %, thereby reducing evaporative heat loss.

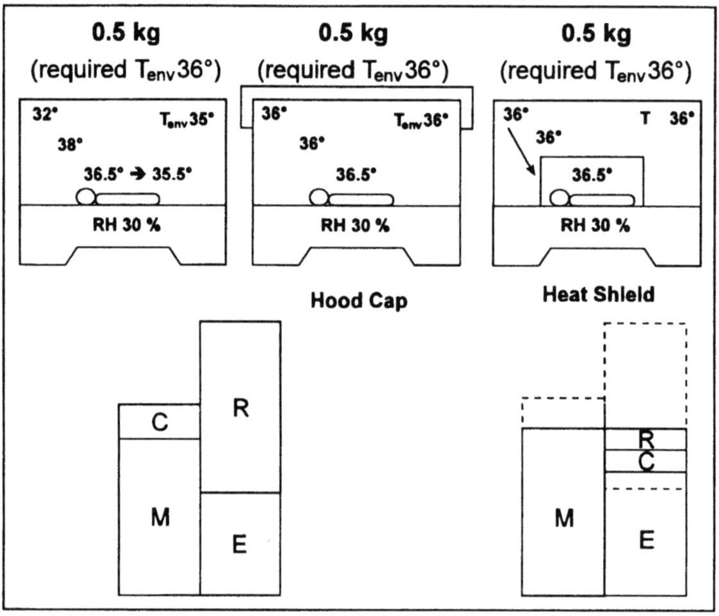

Fig. 12.-4 : This figure illustrates correwction of the heat imbalance of the 0.5 Kg infant by adding a hood cap or heat shield, which increases the wall temperature, thereby reducing radiant heat loss (and incidentally replacing the small gain of heat by convection with small loss).

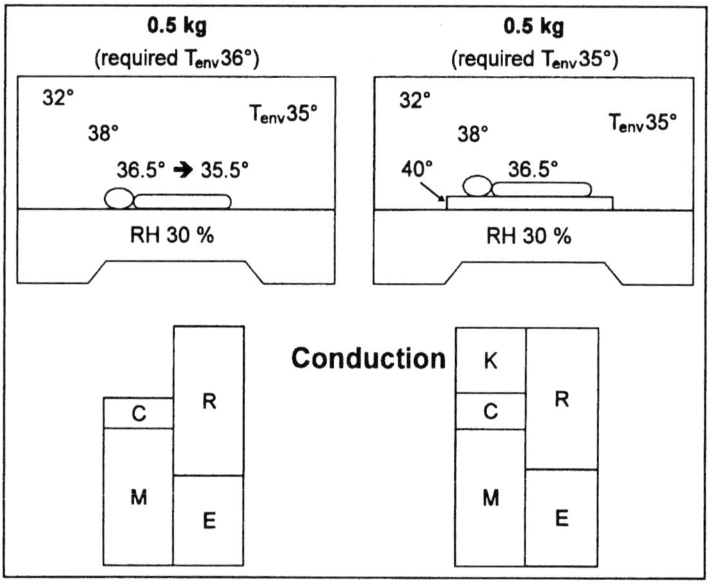

Fig. 12.-5 : This figure illustrates correction of the heat balance of the 0.5 kg infant by adding a heated mattress, thereby producing heat gain by conduction (K).

Modes of Incubator Operation

Incubators can be operated to maintain either a constant air temperature or a constant skin temperature. Either air or skin temperature control can be used with most infants. If the controlling air or skin temperature is carefully selected in the attempt to provide neutral thermal conditions, the mode of control appears to have impact on the infant's metabolic heat production or on the mean body and environmental temperatures [17, 18]. However, greater stability of the air temperature is provided with air temperature control than with skin temperature control [18, 19].

The physiological significance of the more constant air temperature with air temperature control is unknown. On one hand, it may prevent the increase in apnea described in susceptible preterm infants during periods of rising incubator air temperature [20]. On the other hand, the greater variability in incubator air temperature seen during skin temperature control may facilitate the development of intrinsic thermoregulatory control mechanisms. Research is needed to clarify the impact of variability in environmental temperature on the outcome of preterm infants. In practical use, air temperature control offers several advantages over skin temperature control. First, air temperature control makes it possible to maintain air temperature near the maximum without causing the incubator heater to turn off. Second, air temperature control facilitates the use of body temperature to detect a febrile response to infection. Third, air temperature control does not depend on the contact of a sensing probe with the skin.

Monitoring Body Temperature

The task of monitoring the body temperature of preterm infants is an important one. Normally, skin temperature is monitored continuously, whereas core (rectal or axillary) temperature is checked only intermittently. Continuous monitoring of both core (rectal or esophageal) and skin temperatures and their gradient provides information on the infant's vasomotor tone and therefore on his level of thermal stress. However, continuous monitoring of core temperature provides significant practical challenges. Even if these challenges can be overcome, more research is needed to explore the practical value of monitoring the internal thermal gradient. Another potentially fruitful area for research is techniques for the non-invasive monitoring of skin temperature. Successful development of non-invasive temperature monitoring techniques would avoid the artifacts often seen with the use of thermistor probes attached to the skin.

Summary

Incubators have been used since the mid-nineteenth century to provide a supportive thermal environment for sick and preterm infants. In recent decades, open radiant warmers have also been used for this purpose. When compared to radiant warmers, incubators provide partial protection from environmental noise and allow lower insensible water loss. On the other hand, incubators may require special adaptations to provide sufficient thermal protection for the smallest, youngest preterm infants. Moreover, incubators are subject to perturbation of the thermal environment whenever the door is opened for access to the infant. The thermal protection provided by incubators can be enhanced by reducing body heat loss (or increasing gain) by any of the four routes of heat exchange: convection, evaporation, radiation or conduction.

The heat output of an incubator can be regulated to control either air or skin temperature. Air temperature control modulates the variability in air temperature that occurs with skin temperature control, but the physiological value of this modulation is unclear. Maintaining a very stable air temperature may reduce the incidence of apnea of prematurity among susceptible infants but may also delay the maturation of intrinsic thermoregulatory control mechanisms. Further study of this issue is needed.

Additional research is also needed in two other areas. First, more information is needed on the possible use of the internal thermal gradient (core-skin gradient) in the thermal care of preterm infants. Second, there is a need for development of non-invasive methods of measuring body temperature.

Implications for Practice

1. Incubators to be used with small preterm infants should have the capability to manually override the upper limit of air temperature (now typically 37 to 38 °C).

2. Added humidity should be considered for small, young preterm infants in incubators.

3. Plastic heat shields are an inexpensive and effective way to reduce radiant heat losses if more thermal support is needed. Active auxiliary heating devices with separate controls are less desirable because they risk overheating.

4. Incubators manufacturers should strive to further reduce noise levels within incubators.

5. Whenever possible, procedures should be performed through the portholes to minimize disruption of the thermal equilibrium.

6. Either incubator or radiant warmer can be used effectively for most infants.

Recommendations for Future Research

1. More information is needed on the impact of environmental temperature fluctuations on the outcome of preterm infants.

2. More research is needed to assess the possible role of monitoring the internal temperature gradient in the thermal care of preterm infants.

3. Improved methods of monitoring body temperature are needed, methods that are less invasive and less prone to artifact.

4. Incubator manufacturers should modify control software algorithms to improve the incubator's response to thermal perturbation.

References:

1. Denuce (1857) Berceau incubateur pour les enfants nés avant terme. J Med Bordeaux 2:273.

2. Cone TE Jr (1981) The first published report of an incubator for use in the care of the premature infant. Am J Dis Child 135:658-660.

3. Bell EF (1983) Infant incubators and radiant warmers. Early Hum Dev 8:351-375.

4. Glatzl-Hawlik MA, Bell EF (1992) Environmental temperature control. In:Polin RA, Fox WW (eds) Fetal and Neonatal Physiology. WB Saunders Co, Philadelphia, pp 515-526.

5. Williams PR, Oh W (1974) Effects of radiant warmer on insensible water loss in newborn infants. Am J Dis Child 128:511-514.

6. Wu PYK, Hodgman JE (1974) Insensible water loss in preterm infants: changes with postnatal development and non-ionizing radiant energy. Pediatrics 54:704-712.

7. Bell EF, Weinstein MR, Oh W (1980) Heat balance in premature infants: comparative effects of convectively heated incubator and radiant warmer, with and without plastic heat shield. J Pediatr 96:460-465.

8. Bell EF, Oh W (1994) Fluid and electrolyte management. In: Avery GB, Fletcher MA, MacDonald MG (eds) Neonatology. Pathophysiology and Management of the Newborn. JB Lippincott Co, Philadelphia, pp 312-329.

9. Wheldon AE, Rutter N (1982) The heat balance of small babies nursed in incubators and under radiant warmers. Early Hum Dev 6:131-143.

10. Association for the Advancement of Medical Instrumentation (1991) American National Standard. Infant Incubators. AAMI, Arlington, Virginia.

11. Hammarlund K, Nilsson GE, Öberg PÅ, Sedin G (1977) Transepidermal water loss in newborn infants. I. Relation to ambient humidity and site of measurement and estimation of total transepidermal water loss. Acta Paediatr Scand 66:553-562.

12. Sauer PJJ, Dane HJ, Visser HKA (1984) Influence of variations in ambient humidity on insensible water loss and thermoneutral environment of low bith weight infants. Acta Paediatr Scand 73:615-619.

13. Patel CA, Bell EF (1994) Thermal effects of an incubator hood warmer in premature infants. Pediatr Res 35:247A.

14. Hey EN, Mount LE (1967) Heat losses from babies in incubators. Arch Dis Child 42:75-84.

15. Topper WH, Stewart TP (1984) Thermal support for the very-low-birth-weight infant: role of supplemental conductive heat. J Pediatr 105:810-814.

16. Sarman I, Tunell R (1989) Providing warmth for preterm babies by a heated, water filled mattress. Arch Dis Child 64:29-33.

17. Bell EF, Rios GR (1983) Air versus skin temperature servo control of infant incubators. J Pediatr 103:954-959.

18. Chessex P, Blouet S, Vaucher J (1988) Environmental temperature control in very low birth weight infants (less than 1000 grams) cared for in double-walled incubators. J Pediatr 113:373-380.

19. Ducker DA, Lyon AJ, Russell RR, Bass CA, McIntosh N (1985) Incubator temperature control: effects on the very low birthweight infant. Arch Dis Child 60:902-907.

20. Perlstein PH, Edwards HK, Sutherland JM (1970) Apnea in premature infants and incubator-air-temperature changes. N Engl J Med 282:461-466.

Chapter 13

Treatment of Sick Newborns under Radiant Warmers

STEPHEN BAUMGART

Introduction

Maintaining core body temperature of premature newborn infants exposed to a hostile extrauterine environment in the modern intensive care nursery is essential for the survival and growth of these delicate patients [1-5]. The concept of a thermal neutral environment for best nurturing such babies has become widely accepted [6-11]. The thermal neutral environment may be defined as the set of environmental conditions (particularly environmental temperature) that results in the minimum expenditure of the infant's precious metabolic energy reserve to maintain body core temperature within a narrow, homeothermic range (37.2-37.5 °C) [9]. Originally, the thermal neutral environment was described within conventional convection warmed incubators in terms of the optimal air temperature and relative humidity, and the optimal incubator mean wall temperature. Thermal neutral conditions varied for infants of different body weights, and gestational and postnatal ages [9, 10].

Recognizing that environmental and body temperatures may fluctuate, however, the definition of thermal neutral conditions more recently has been translated to incorporate the infant's skin temperature [12-14]. Defined in this way, the thermal neutral environment may be any set of environmental conditions resulting in an optimal skin temperature and corresponding to a minimum in the observed resting metabolic energy expenditure. Empirically, 36.0-36.2 °C anterior abdominal wall skin temperature has been designated as thermally neutral for most premature infants nursed conventionally inside of incubators [12].

However, neonatal intensive care practices now include rescue of critically ill infants in severe cardiorespiratory failure, and very low birth weight infants under 750 grams at birth. Treating such patients is beyond the scope of original recommendations for optimal incubation, and often requires frequent physical access for diagnostic and therapeutic procedures. The maintenance of a thermal neutral temperature despite frequent interruption of the conventional incubator's steady state environment (by opening the hand ports or side walls) has become a major problem [15]. Considerable fluctuation in environmental temperature may occur with these interventions, and a non-steady state, non-thermal neutral environment results.

Original reports by Hey and Katz [10], Agate and Silverman [16], Levinson *et al* [17], and Silverman *et al* [12, 18], and Price *et al* [19], suggested using low energy infrared heat sources to reduce the convective temperature fluctuations experienced within incubator hoods. More recently, high powered infrared warming devices, positioned over open bed platforms and servocontrolled to maintain skin temperature between 36.5 and 37 °C, have been demonstrated as an alternative to conventional enclosed convection warming, and avoid the problem of constant incubator interruption [16-19]. Moreover, radiant warmer beds have recently been shown to provide thermal neutral conditions for the management of critically ill, low birth weight infants during the first two weeks of life [20].

However, some reports criticize routine and prolonged use of radiant warmer beds to nurture premature infants, and suggest that such a strategy incurs increased evaporative water loss from the skin of premature babies [21-28], increased minimal observed metabolic rate [29], and harbors the potential for infrared injury to delicate developing tissues, particularly the anterior structures of the eye [30]. In the sections to follow, the development and application of radiant warming technology between 1970 and 1990 will be reviewed. In particular, the clinical description of partitional calorimetry, the definition of a thermal neutral management strategy, shielding for infants under radiant warmers from convection and evaporation, and exploring possible detrimental effects of infrared energy on the developing human eye will be discussed.

A General Description of open Radiant Warmer Beds

Open radiant warmer beds generally comprise a patient bed platform located at waist level with sufficient area for delivering neonatal intensive care, and openly exposed to the nursery's room air. The performance of neonatal nursing and respiratory care procedures (blood sampling, parenteral nutrition and arterial catheter care, and respiratory tubing changes), technical evaluations (radiography, echocardiography and cranial ultrasound), and minor surgical procedures is thus facilitated by open access without the encumbrance of walls, door panels or hand ports.

Suspended on an armature approximately 80-100 cm over the baby is a radiant warming module. Usually this module contains an electronic warming element comprised of a wire heating coil contained within a quartz tube envelope [30, 31]. This element generates radiant heat through resistance to electrical current passed through the wire which causes extreme temperature elevation of the metal alloy to the point where infrared electromagnetic radiation is generated. Absorption of infrared heat from the wire causes secondary warming of the quartz tube envelope to a high, but lesser temperature. Radiation from both the wire heating element and the quartz envelope summated transmits infrared energy directly onto (and into) the baby's skin. Such devices may consume 500-600 watts at maximum electrical power input, and may deliver a net radiant power density output (at peak performance) between 35-50 milliwatts per square centimeter (mW/cm^2, *radiant power density*) surface area irradiated at the bed platform level. The heating element is often encased within a polished aluminium reflector capable of directing radiant heat delivery onto the bed platform.

Power generated from the heating element and reflector apparatus is usually governed by an electronic servocontrol module which varies electrical power input in proportional response to measured infant skin temperature determined by a sensitive electronic thermistor probe usually taped to the infant's abdomen [33]. The temperature probe is a rapid response, passive resistance thermistor connected through a Wheatstone bridge to the servocontrol module. Second by second, the measured infant temperature modulates warmer output by either proportional, differential, or integrated electronic logic. For the most part, other more complex (or more responsive) logical (algorithmic) systems are currently considered experimental. Most clinicians agree, that for the nurturance of premature infants under such devices in the first few weeks of life, an anterior abdominal wall skin temperature servocontrolled to maintain (as consistently as possible) a temperature somewhere between 36.0 and 37.2 °C (depending on infant size, gestation and postnatal age) is sufficient to maintain infant core homeothermy [20, 29, 34-39].

Partitional Calorimetry under Radiant Warmers

The first reports of low energy radiant warming (used in conjunction with conventional, convection warmed incubators) constitute classic descriptions of partitional calorimetry. Infrared heat was used in these studies to provide an incubator wall temperature consistent enough for the definition of the thermal neutral environment over a wide range of premature infant weights, and of various gestational and postnatal ages [10, 16]. Often, the authors of these studies provide a footnote stating that incubator air temperature selection should be adjusted for the nursery's room air (and room wall) temperature, since most commercially available devices do not provide heated wall panels. Partitional calorimetry, however, performed for infants nurtured on open bed platforms under high power radiant heating devices was difficult to describe with the early introduction of these devices, and was not widely understood until recently.

Partitional calorimetry is performed by describing the infant's metabolic rate of heat generation and body temperature maintenance within the context of the physical parameters of heat exchange within the baby's environment: convection, conduction, evaporation and radiation [40-43]. These parameters may be summarized by the general metabolic equation as follows:

$$Q_{metabolic} + Q_{stored} = Q_{convection} + Q_{conduction} + Q_{evaporation} + Q_{radiation},$$

where $Q_{metabolic}$ is the infant's metabolic rate of heat generation, Q_{stored} is the infant's rate of heat storage based on the increase of the infant's body temperature over time (usually zero at thermal equilibrium, or negative if heat loss results in a reduction of body temperature), $Q_{convection}$ is the rate of heat exchange between the infant's skin and the surrounding air (heat loss assumes a positive value, conventionally, whereas heat gain assumes a negative value), $Q_{conduction}$ is the rate of heat exchange (positive for loss, negative for gain) from the infant's skin in contact with solid surfaces, $Q_{evaporation}$ (the latent heat of water's state change from a liquid to a gaseous phase) is the rate of either heat loss from evaporation from the infant's skin and respiratory tract (positive value), or conversely from condensation of water vapor from the environment onto the infant's airways or skin (negative value), and $Q_{radiation}$ is the rate of heat exchange due to electromagnetic transmission of infrared energy either away from (positive) or into (negative) the infant. Each of these parameters may usually be expressed in kcal/kg/hour, in watts/m^2, or in kilojoules/kg/day.

In the classic thermal equilibrium described by human adult physiologists, convection, conduction, evaporation, and radiation are considered to be moderate heat losses occurring under normal temperate environmental conditions. Such a description has also been reported consistently for infants nurtured within conventional convection warmed incubators—usually, incubator air and wall temperatures are slightly less than the infant's skin, and relative humidity is less than 100 % saturated. However, for infants nurtured under a variety of different thermal conditions (*e.g.* high powered radiant warmers in particular), these avenues of heat loss may be altered quantitatively, in some instances contributing to substantial heat gain. Table 13.-1 summarizes the four avenues of physical heat exchange for describing heat losses (or heat gains) produced by incubation devices with the mnemonic "earth, air, fire and water" [39].

Tab. 13.-1 : Four avenues of physical heat transfer. (With permission, C.V. Mosby & Co., In Intensive care of the Fetus and Newborn, Spitzer AR ed, Baumgart, *Thermal regulation in the fetus and newborn infant*, 1995)[39]

1. "Earth"	=	Conduction
2. "Air"	=	Convection: a. Natural b. Forced
3. "Fire"	=	Radiation
4. "Water"	=	Evaporation (or condensation)

The earliest descriptions of partitional calorimetry for infants nurtured under high powered radiant warmers were provided by Wheldon and Rutter [44, 45], and by Baumgart [41]. Wheldon and Rutter's first work characterized measurements of conduction, convection, evaporation and radiation in incubators and under radiant warmers in a specially constructed infant manikin [45]. The latter work by Baumgart constituted empirical observations made during the practice of radiant warming in an actual intensive care nursery, and will be the focus of the following section [41]. Partitional calorimetric studies were conducted in ten infants ranging 1.08-1.80 kg and 28-34 weeks gestation at two weeks or less of life. Most infants were recovering from respiratory distress syndrome. Babies were nursed naked and supine on open bed platforms within the intensive care nursery under radiant warmers servocontrolled to maintain anterior abdominal wall skin temperature at three different settings for a period of 3 hours each at 35.5, 36.5, and 37.5 °C. Order of temperatures studied was randomized.

Assessment of Convective Heat Loss, ($Q_{convection}$).

At least four authors in contemporary studies have reported the assessment of convective heat loss in premature neonates using a partitional description for both natural and forced convection [25, 44-47]. We used equations developed from adult studies conducted by Colin and Houdas [48], extrapolated to premature subjects. Parameters affecting the rate of convective heat transfer are elaborated elsewhere in this text.

In order to measure forced convection's effect on heat loss in infants under radiant warmers, air velocity was measured using a hot wire anemometer [25, 41, 44, 45, 47], which quantitates air velocity to as low as 7 cm per second (0.25 km/hour). Air velocity was assumed to affect the infant primarily in vectors traversing from head to foot or from foot to head. Air velocity was integrated over time and mean air velocities determined [25]. Other factors (including the infant's radius of body curvature less than an adult's, the effects of posture and activity, and other turbulence vectors in the nursery) may have resulted in consistent underestimation of this value [47-49].

Assessment of Insensible Water and Evaporative Heat Loss, ($Q_{evaporation}$).

In order to assess the infant's rate of evaporation of water (both transcutaneously and from the respiratory tract), continuous monitoring of the infant's rate of weight change in grams/kg/hour was performed using an electronic metabolic baby scale. Change in body weight over time has been termed insensible weight loss and may be expressed by the following equation:

$$IL = dW_{scale} + (dW_{intake} - dW_{output}),$$

where IL is the net rate of metabolic weight loss in grams/kg/hour,

dW_{scale} is the rate of change in weight of the baby scale over the study period (positive for loss, negative for gain) in grams/kg/hr, and (dW_{intake} - dW_{output}) is the difference in the rates of weight change between the number of grams of fluid (nutritional) intake infused parenterally or enterally, and fluid (urine flow, or blood sampled, or stool excreted, *etc*) outflow in grams/kg/hour [26, 42]. Net metabolic weight loss (IL) has been related to insensible water loss (IWL) and the infant's metabolic rates of gas exchange by the Isenschmid equation which accounts for the difference in weights between the metabolic rate of oxygen consumption and the rate of exhalation of carbon dioxide measured by indirect calorimetry [50]. A maximum effect of respiratory gas exchange on insensible water loss calculations may comprise 10 to 15 % of this measurement. Calculation of the rate of heat loss through evaporation is obtained by multiplying insensible water loss times 0.58 kilocalories per milliliter of water evaporated (at 37 °C skin temperature) to yield the rate of evaporative heat loss in kilocalories/kg/hr.

In our studies, scale calibration was performed after tubing and wires were fixed to the platform and warmer bed, and 3 hours equilibration time for thermal and respiratory humidification apparatus equilibration. Accuracy was demonstrated on a polygraph record during each study by placing a series of calibrated weights between 0.5 and 5 grams onto and off of the scale platform, and observing the arc subtended on the polygraph record as shown in figure 13.-1 [42].

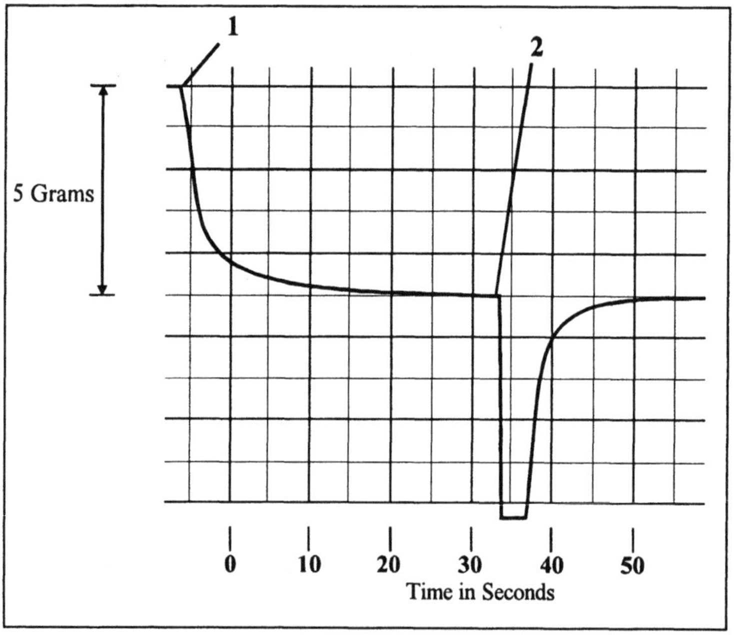

Fig. 13.-1 : Weight change tracing from the Potter electronic baby scale. (With permission, S. Karger Publishing Corporation, Baumgart, Biology of the Neonate, 1990; 57: 10)[42]

When 5 grams is removed at point 1 on this figure, 25 mm is subtended on the polygraph by the scale, returning to its 0 point at the middle of the polygraph record. Scale error encountered by this technique of continuous rate of weight change monitoring was estimated to be less than 0.5 grams/kg/hour.

This technique accounted for several sources of scale error encountered by previous investigators when measuring evaporative heat loss using this method of determining metabolic weight change including: [1] load drag on the scale platform from wires and tubing attached to the infant, [2] scale hysteresis (memory effect, figure 13.-1, point 2), [3] condensation of respiratory gas humidification within ventilator circuits [51], [4] the specific gravity of fluids (both being consumed and eliminated) during the course of the study, and [5] thermal expansion (or contraction) of the stainless steel lever load mechanism of the scale (Figure 13.-2) [42, 52, 53].

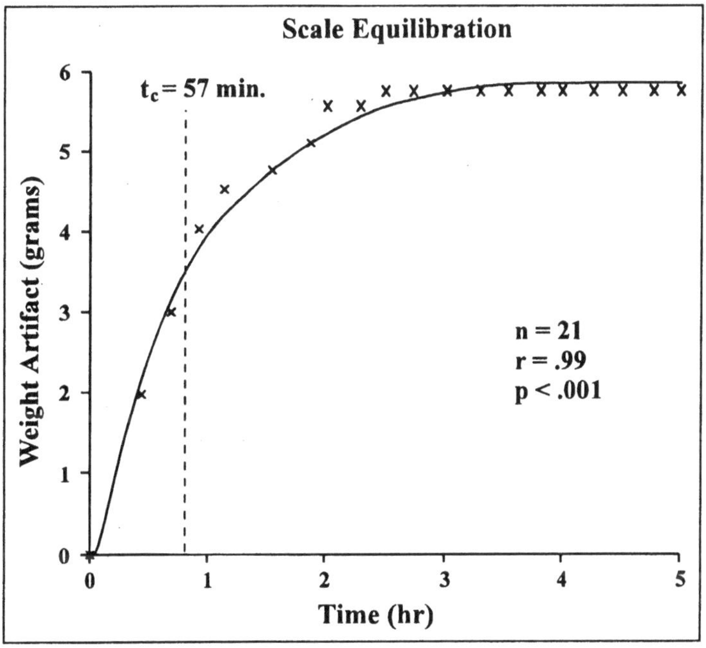

Fig. 13.-2 : Weight change artifact observed as the Potter scale reaches thermal equilibrium under a radiant warmer at half power. (With permission, S. Karger Publishing Corporation, Baumgart, Biology of the Neonate, 1990; 57: 10)[42]

Determination of Metabolic Rate of Heat Production, ($Q_{metabolic}$).

Metabolic rate of heat production was monitored by indirect calorimetry in intubated infants on CPAP as depicted in figure 13.-3 [41]. Oxygen consumption and carbon dioxide production were measured in ml/kg/min according to the relationship described by Lusk [54]. An open circuit technique was used to monitor these rates of gas exchange in non-intubated subjects continuously as reported by Schultz [55, 56]. After gas collection and dessication using a dry ice condenser and calcium sulfate canisters, F_IO_2, F_EO_2 and F_ECO_2 were measured using a dual channel paramagnetic oxygen analyzer and an infrared CO_2 capnometer. Inspired carbon dioxide content was considered to be 0. Results were corrected to standard temperature and pressure (dry) conditions. Computer sampled recordings were made for gas signals at 1 Hz. Gravimetric gas standards were used to calibrate this device and the ratio between O_2 and CO_2 was calibrated by alcohol lamp.

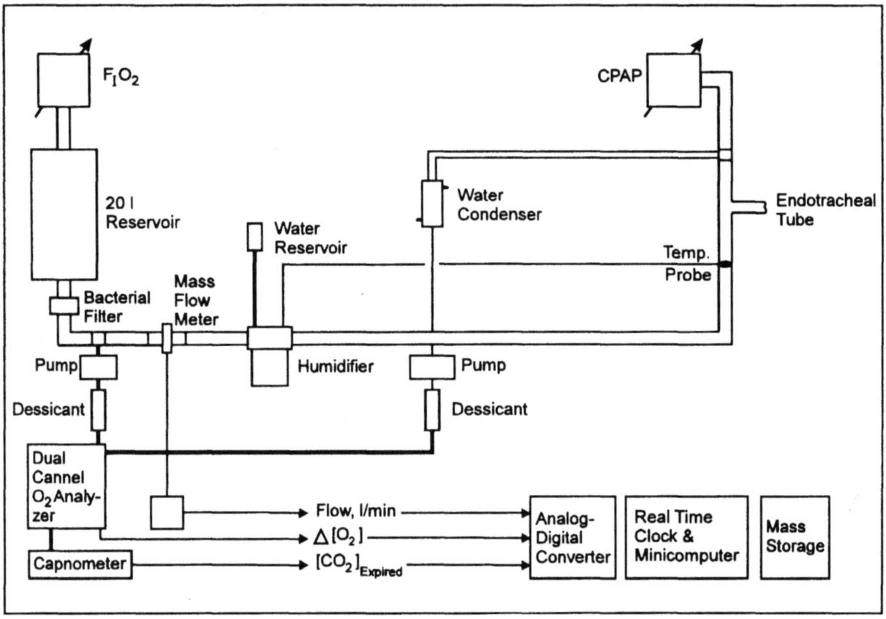

Fig. 13.-3 : Diagram of an indirect calorimeter configured for open circuit measurements in infants intubated endotracheally. (With permission, American Academy of Pediatrics, Baumgart, Pediatrics, 1985; 75: 89)[41]

For infants intubated and receiving continuous positive airway pressure, the bias introduced in assessment of metabolic rate is primarily through endotracheal tube air leak. Such a bias would result in consistent underestimation of oxygen consumption. Conversion of oxygen consumption into metabolic rate of heat production was adjusted based on the fuel mixture burned (fats, carbohydrates, and proteins) according to the respiratory quotient as cited previously [54]. $Q_{metabolic}$ was expressed in kcals/kg/hr. Rather than use minimal observed metabolic rate (MOMR), metabolic rate over the entire study period (including periods of peak activity) was integrated. Total metabolic rate of heat generation was thus assessed [55, 56].

Partitional Calorimetry at Different Servocontrol Temperatures, and Calculating Radiant Heat Needed to Balance the Partition, ($Q_{radiant\ need}$).

Partitional calorimetry performed at three different servocontrol temperatures (35.5, 36.5 and 37.5 °C) is shown in figure 13.-4 [41]. A complete partition at each temperature is described by each cluster of histograms. Heat losses are represented on the left side of this cluster and heat gains at the right. As seen in this figure, the infant's rate of convective heat loss increased as the gradient between the skin and the air temperature increased under the radiant warmer. Moreover, convective heat loss comprised the single largest avenue for heat loss that infants experienced during radiant heating. Ambient nursery air temperature in general ranged from 8-10 °C cooler than skin temperature. Evaporative heat loss may have increased slightly from the coolest to the warmest servocontrol temperature (*ie* from lowest to highest radiant power density delivered to the infant), however, this effect was not significantly different statistically. The histogram bar at the right of each cluster demonstrates the infant's rate of metabolism which diminished with increasing skin temperature, but did not vary significantly.

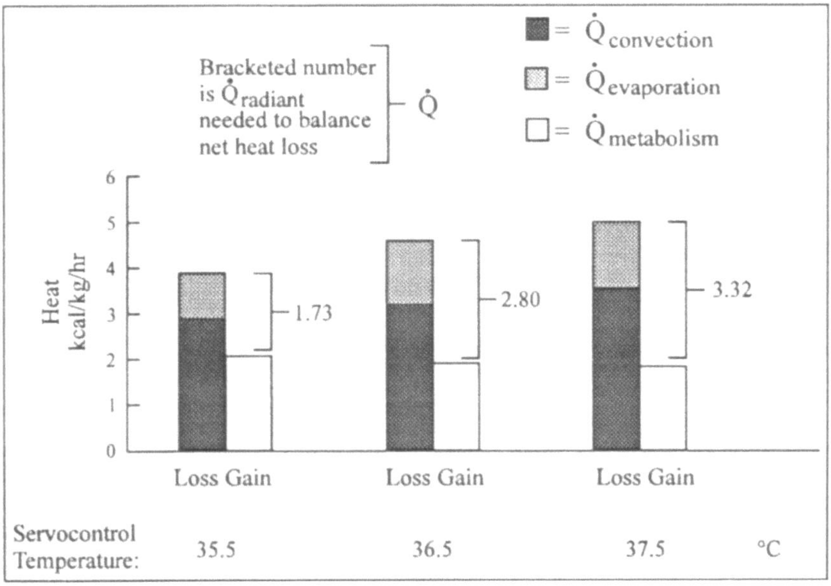

Fig. 13.-4 : Partitional calorimetry demonstrating physical parameters of heat loss or heat gain, and metabolic rate of heat production in premature infants nursed naked and supine under radiant warmers set to servocontrol anterior abdominal wall skin temperature at 35.5, 36.5 or 37.5 °C. (With permission, American Academy of Pediatrics, Baumgart, Pediatrics, 1985; 75: 89)[41]

A tendency was observed for metabolic rate to be somewhat less at higher temperatures than at the lowest temperature by about 15 % of metabolism. The radiant heat needed to balance the partition and thus maintain infant homeothermy ($Q_{radiant\ need}$), was calculated from the partition and is shown in figure 13.-4 by the bracketed numbers. $Q_{radiant\ need}$ ranged from 1.73 kcal/kg/hr at 35.5 °C servocontrol temperature to 3.32 kcals/kg/hr at the highest temperature.

Our conclusions from this study are that convection is the predominant route of heat loss and radiant heat gain from the warmer is the predominant route of heat gain in determining the premature infant's thermal equilibrium. In contrast to a conventional convection warmed incubator, radiant heat replacement is quantitatively more important in maintaining thermal equilibrium than the infant's own metabolic rate of heat production. In fact, the infant's metabolism is overshadowed at higher temperatures by the radiant warmer. As a final observation, insensible water loss did not increase significantly with increasing radiant energy delivery.

A Direct Assessment of Radiant Warmer Heat Delivery, ($Q_{radiant\ delivery}$).

In addition to simply calculating radiant heat need from partitional calorimetry, direct assessment of radiant warmer heat delivery was made quantitatively during these studies using a thermopile transducer, (Eppley Model E6) [31, 32]. This device was placed at bed level approximating the position of the infant's abdomen. This thermopile is a Koblentz type device comprised of a 1 cm^2 inverted black-body cone which receives radiant heat within a 70° incident arc. Fifty copper/constantan junctions convert infrared electromagnetic radiation absorbed by the black, metal oxide surface of the inverted cone into an electrical potential which varies proportionally with the incident radiant heat. Radiant power density is measured by this device (and calibrated to standard radiant heat sources) in milliwatts per square centimeter (mW/cm^2).

Radiant power density measured using the thermopile was compared to electronic power input to the radiant warmer measured by a root/mean square, true value watt meter. As electrical power demand varied (governed by the infant's servocontrolled thermistor) radiant power density delivered to bed level was thereby monitored continuously. A radiant power density map of the warmer bed platform at the level of the infant is shown in figure 13.-5 [25]. As seen from this diagram, the infant's mattress comprises a radiant power density grid which is cooler at the foot, and warmer at the head. Nevertheless, where the infant resides within the central highlighted region of the bed's surface, radiant warmer power density delivered to bed level is relatively homogenous (within 85 % of peak value delivered). Outside of this central area, radiant power density diminishes rapidly, particularly at the bed's foot.

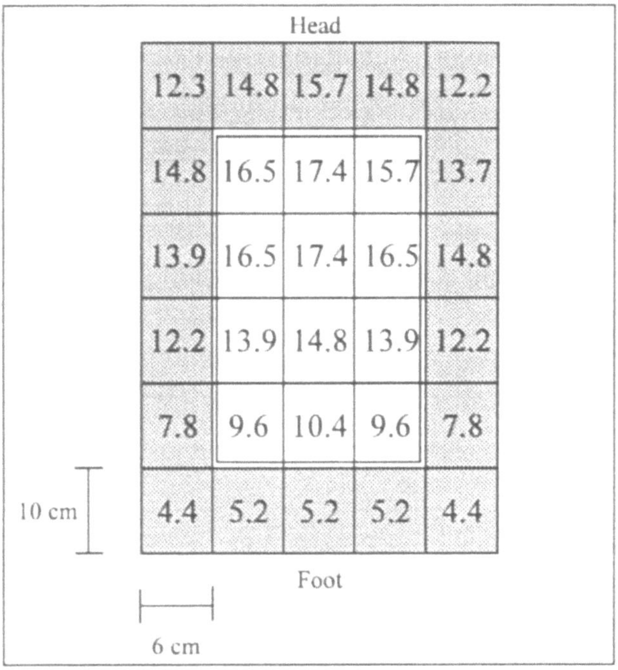

Fig. 13.-5 : Map of radiant power density received (mW/cm^2 surface area irradiated) at bed level by infants nurtured under radiant warmers at about half power. Baby usually resides within the central highlighted area. (With permission, Mosby-Year Book Inc., Baumgart *et al*, Journal of Pediatrics, 1981; 99: 948)[25]

To calculate net radiant heat delivery to the infant over the baby's entire body surface area exposed to the radiant warming element overhead, we modelled the infant as a series of cylinders, demonstrated simplistically in figure 13.-6 [40]. Assuming a plane radiant heat source generated by the points along the heating element's reflector system, we viewed the infant casting a shadow in rectangular form over the bed surface underneath, thus interrupting a rectangle of homogenous incident radiant power density. By simple geometric calculation and summation, radiant energy thus received by the infant was converted from mW/cm^2 (thermopile irradiance), and an infant's body surface area and weight according to the following formula:

$$Q_{\text{radiant delivery}} = (3600/3184)(0.3183)(RPD/1000)(SA/W),$$

where RPD is the radiant power density received at bed level (center) in mW/cm^2 divided by the constant 1000 watts/milliwatt, (SA) is the baby's body surface area calculated from the infant's weight and length using a neonatal nomogram [57], and the constant $1/\pi = 0.3183$ corresponds to the projection factor of radiant power density distributed from a plane heat source onto a curved surface, (W) is the infant's body weight in kilograms [26].

The factors 3600 seconds per hour divided by 3184 joules per kcal convert watts into kcal/hour. $Q_{radiant\ delivery}$ is then expressed in kcals/kg/hour [26].

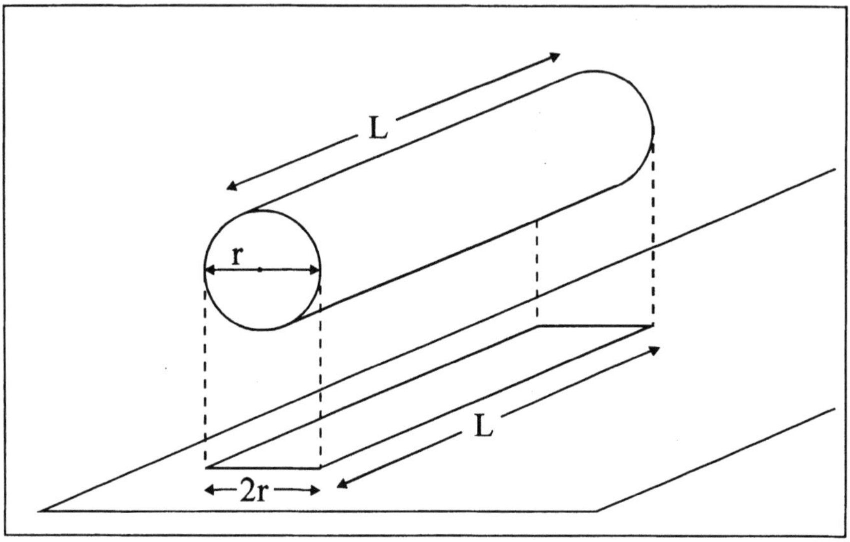

Fig. 13.-6 : Cylindrical model for infant torso, extremities and head used to calculate total incident irradiance absorbed by babies exposed to a radiant heat source overhead. The cylinder projects a planar shadow onto the bed surface. (With permission, Baumgart, Journal of the California Perinatal Association, 1984; IV(2): 15)[40]

Radiant Heat Need ($Q_{radiant\ need}$) *v* Radiant Heat Delivered ($Q_{radiant\ delivery}$).

The comparison between radiant heat need calculated from partitional calorimetry, and the radiant heat actually delivered by the servocontrolled radiant warmer as measured by the thermopile technique described above is shown in figure 13.-7 [41]. Out of 31 observations in 10 subjects, radiant heat delivery correlated to the radiant heat need in a linear fashion. However, apparent radiant heat delivered was greater than radiant heat need in almost every case. The difference between these two observations led us to believe that the infant was experiencing heat loss in excess of that represented by convection and evaporation. Two possible avenues were considered to balance this apparent difference: Conductive heat loss ($Q_{conduction}$), previously disregarded since infants were managed on an insulating foam rubber mattress, or radiant heat loss ($Q_{radiant\ loss}$) to the infant's sides which were not directly facing the radiant heating element.

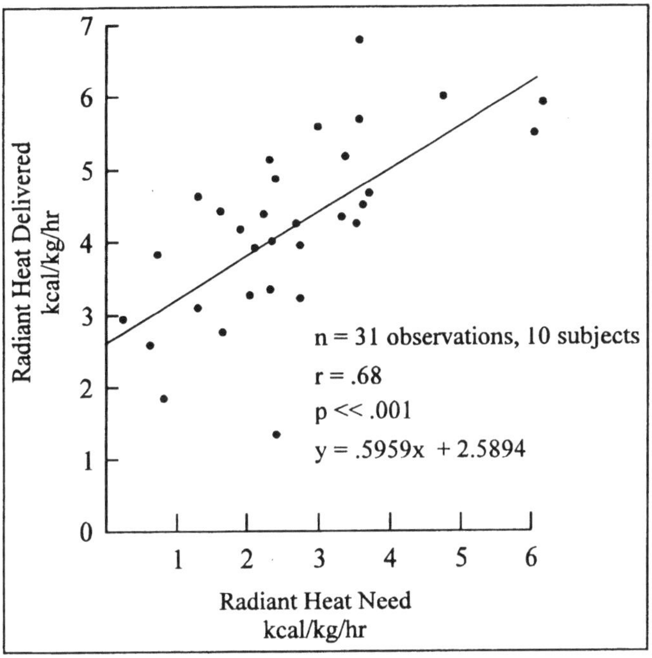

Fig. 13.-7 : Graph demonstrating calculated radiant heat needed to balance the partitional calorimetry demonstrated at 36.5 °C servocontrol set-point temperature in figure 13.-4 *versus* the measured radiant power delivered by the radiant warmer. A discrepancy is demonstrated where actual power delivery generally exceeds heat need calculated. (With permission, American Academy of Pediatrics, Baumgart, Pediatrics, 1985; 75: 89)[41]

Toward Describing Complete Partitional Calorimety under Radiant Warmers.

To quantitate conductive heat loss to the insulated mattress surface we measured the infant's bed platform (mattress) temperature *versus* the infant's interscapular skin temperature, and estimated partial body surface area contact with the bed surface [42, 46]. To evaluate the possibility of radiant heat loss experienced by infants under radiant warmers, $Q_{radiant\ loss}$ was estimated by considering the following equation (expressions previously defined above):

$$Q_{radiant\ loss} = Q_{radiant\ delivery} - Q_{radiant\ need}.$$

Using these radiant heat exchange calculations, and measuring conduction, an adjusted partitional calorimetry for a more complete description of the radiant warming environment at a servocontrolled skin temperature of 36.5 °C is demonstrated in figure 13.-8 [42]. As seen at the left, heat loss is represented by convection 64 %, evaporation 30 %, conduction 4 % and radiant heat loss to the infant's sides by 2%. Heat gain is represented by direct radiant energy delivery measured by thermopile at 58 %, and by metabolic rate of heat production by the infant at 42 %. This comprises the most complete measurement of partitional calorimetry for premature human neonates nurtured under a radiant warmer and parallels closely the modelled heat partition demonstrated by Wheldon and Rutter previously [44, 45]. We conclude from these studies that although there may be some component of radiant heat loss felt at the sides of infants nurtured under radiant warmers, this component contributes little to the thermal equilibrium established between larger components (particularly convection and radiant heat delivery).

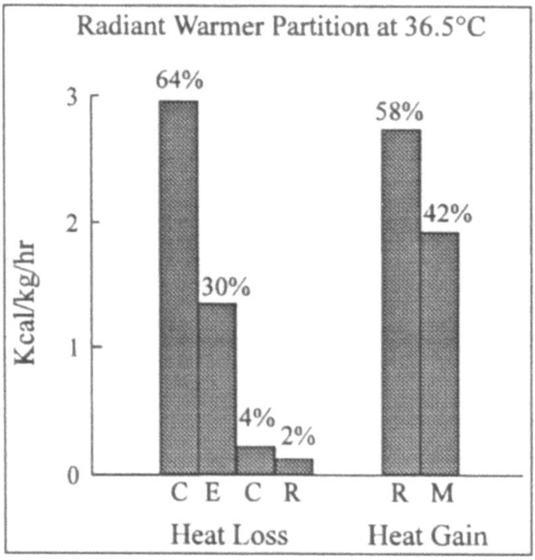

Fig. 13.-8 : Complete description of partitional calorimetry including radiant heat loss and conductive heat loss for infants nurtured naked and supine under radiant warmers servo-controlled to maintain anterior abdominal wall skin temperature at 36.5 °C. (Adapted with permission, S. Karger Publishers, Baumgart, Biology of the Neonate, 1990; 57: 10)[42]

Radiant Temperature and Temperature Homeothermy.

Nevertheless, the radiant temperature experienced by the infant's skin may not be homogenous with respect to radiant environmental temperature. To demonstrate this effect, the thermopile was again positioned at bed level and moved through an arc from 0 to 180°. As shown in figure 13.-9, the direction of heat delivery from the warmer's heating element is overhead, and the infant's head, feet and sides are as oriented laterally on the diagram [42]. Thermopile measurements demonstrated the greatest density of radiant power delivery at 90° (or head-on). As the radius of curvature demonstrates, radiant heat delivery falls off to the infant's sides approaching 0 (with the thermopile referenced to the room wall temperature of the nursery). Since the nursery's wall and air temperatures are at equilibrium between 25-27 °C, radiant heat loss would be expected to result from the temperature gradient between the infant's skin at 36-36.5 °C, and the surrounding cooler walls of the nursery.

Following from this observation, the infant's radiant environment may be described by a conceptual mean radiant temperature ($T_{radiant}$), which would comprise a theoretical mean wall temperature required for a warmed enclosure surrounding an infant nursed on an open bed platform at ambient room air temperature. $T_{radiant}$ was calculated using the Stefan-Boltzmann relationship as follows:

$$T_{radiant} = [T_{mean\ skin}^4 - (Q_{radiant\ need})(W)/(\sigma)(\varepsilon_{skin})(\varepsilon_{environment})(A)(F)]^{-0.25},$$

where σ is the Stefan-Boltzmann constant, $4.93 \times 10^{(-8)}$ kcals/hr/m^2/^{o}K^{-4}, ε_{skin} is the radiant emissivity of the infant's skin (assumed to be one since infants are considered black body irradiators), $\varepsilon_{environment}$ is the emissivity of the complex radiant environment surrounding the baby (assumed to be near one for the purpose of this calculation) [42]. Overestimation in emissivity may result in modest overestimation of the radiant temperature. All temperatures expressed in ^{o}K are converted from °C by adding 273. Area (A) is the infant's body surface area [57], (F) is the fraction of effective radiating body surface area exposed (assumed to be 0.5 after the estimation of Bell and Rios) [46], and (W) is the infant's weight in kgs.

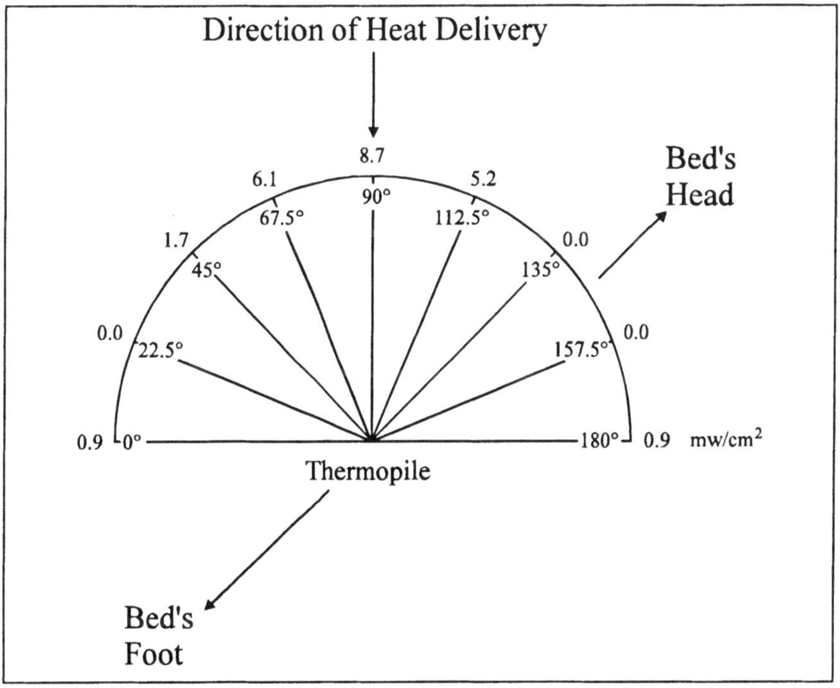

Fig. 13.-9 : Radiant heat delivery measured by a thermopile moved through a 180° arc at bed level under a radiant warmer at about half power. Irradiance diminishes as the thermopile faces the walls to the sides of the warmer bed platform rather than directly facing the warming element. (With permission, S. Karger Publishing Corporation, Baumgart, Biology of the Neonate, 1990; 57: 10)(42)

The $T_{radiant}$ calculated from these data for infants nurtured under radiant warmers servocontrolled to 36.5 °C was approximately 45 °C. The difference between this temperature and the nursery walls (25-27 °C), comprises the contribution of the overhead radiant warming element to the infant's exposed surface. The infant's skin probably senses this disparity, particularly at the infant's sides.

Radiant Warmers and Insensible Water Loss in Prematures

One of the major problems identified with use of open radiant warmer beds for incubating premature infants was the observation of increased insensible water loss, (Table 13.-2) [22, 24, 26, 58, 59]. Particularly when compared to humidified incubators, water loss under radiant warmers was thought to be increased due to a direct effect of radiant energy on the skin [24]. In this study, Bell and co-workers demonstrated an increase in insensible water loss from 1.58 to 2.43 ml/kg/hr when infants were moved from a convection warmed incubator onto an open radiant warmer bed platform.

Tab. 13.-2 : Insensible water loss studies reporting values for babies nursed in conventional convection warmed incubators versus under radiant warmers.
(With permission, W.B. Saunders & Co., Costarino and Baumgart, Pediatric Clinics of North America, 1986; 33)

Reference	Infant Weights, [kg]	Incubation Device	IWL, [ml/kg/h]
Fanaroff 1972 [58]	0.70-1.25	Incubator	3.45±0.67
	1.25-1.80	Incubator	1.41±0.71
Wu 1974 [22]	<1.00	Incubator	2.68±0.18
	1.00-1.25	Incubator	2.32±0.31
	1.26-1.50	Incubator	1.60±0.30
	1.51-1.75	Incubator	0.92±0.25
	1.76-2.00	Incubator	0.71±0.15
	<1.50	Radiant Warmer	2.45±0.40
	>1.50	Radiant Warmer	1.49±0.40
Williams 1974 [59]	3.10±0.13	Incubator	0.53±0.05
	3.24±0.97	Radiant Warmer	1.08±0.12
Bell 1979 [24]	0.79-1.31	Incubator	1.58±0.26
		Radiant Warmer	2.43±0.24
Baumgart 1981 [26]	0.66-1.00	Radiant Warmer	5.25±0.77
	1.01-1.50	Radiant Warmer	2.66±0.20
	1.51-2.00	Radiant Warmer	0.52±0.01

When the same infants were additionally supplied phototherapy (felt by the authors to constitute an additional source of radiant energy), insensible water loss increased from 2.43 to 3.51 ml/kg/hr. Thereafter, addition of an additional phototherapy unit brought insensible water loss to 3.54 ml/kg/hr. The authors concluded that additional application of radiant energy sources incrementally increased insensible water loss at the baby's skin-air interface, and they attributed these water losses to the effect of additional radiant heating of water in the skin.

In our own investigations of radiant energy's impact on insensible water loss, little incremental increase in insensible water loss was observed in infants treated with increasing levels of infrared power density [41, 61, 62]. We investigated the effect of phototherapy on the quantity of radiant heat delivered to infants under radiant warmers and on insensible water loss as demonstrated in figure 13.-10 [60]. In the left set of histogram bars, servocontrolled radiant warmer power with and without phototherapy demonstrates a similar net radiant power density at bed level under these two conditions. The servocontroller on the radiant warmer exactly compensates for the incident radiant energy at the baby's skin resulting from exposure to the phototherapy unit. Nevertheless, the observation of increased insensible water loss with the addition of phototherapy is shown by the histogram bars at the right of this figure.

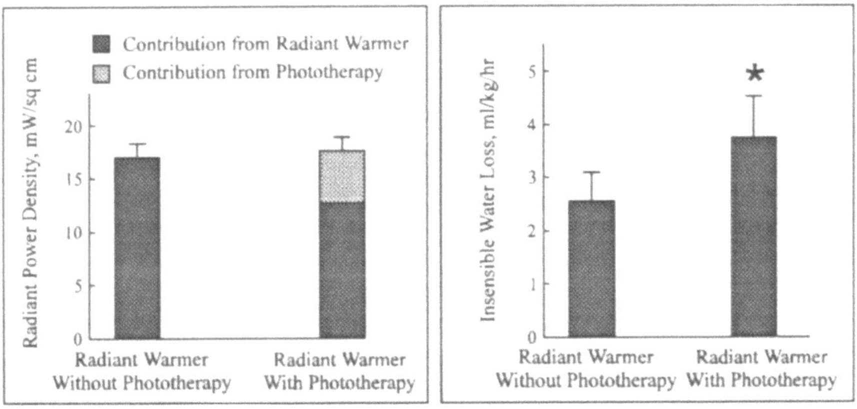

Fig. 13.-10 : Phototherapy exposure results in an increase in insensible water loss in infants nurtured under radiant warmers, however, the warmer's servocontrol mechanism exactly compensates for the excess irradiance delivered by the phototherapy unit, render radiant heat exposure the same with or without phototherapy. (Adapted with permission, American Medical Association, Engle *et al*, American Journal Diseases in Children, 1981; 135:516)[60]

Therefore, we speculated that perhaps not the infrared heat exposure but the intense visible light was the major effector increasing insensible water loss. An increase in skin blood flow (and therefore skin water content) may have occurred with exposure to visible light contributing to transcutaneous evaporation. Alternative explanations for this phenomenon include increased respiratory rate or increased rate of stooling, neither of which we observed in our study. We are left with the question whether radiant energy *per se* or some other environmental factor may contribute to the observed increased insensible water loss observed by Bell *et al* under radiant warmers.

The Role of Convection in Evaporation under Radiant Warmers.

We further evaluated the impact of environmental manipulation on thermal equilibrium (in particular, on insensible water loss) in premature neonates under radiant warmers in a study of three separate environments [25]. For the first environment (control), an infant was nursed naked and supine under a radiant warmer on an open bed platform. For the second, the same infant was nurtured with clear plastic walls surrounding the open bed platform. Many of the studies previously conducted which demonstrated increased insensible water loss under radiant warmers used the same metabolic baby scale described above. By placing the baby on this scale (11-12 cm above the bed platform), the warmer bed's side walls are obviated. From our previous experience measuring forced convection, we had noted that opening and closing nursery doors, movement of personnel past the bed platform, and cycling of air ventilation within the intensive care unit generated air turbulence which affected babies on the scale. Therefore, we built the walled chamber above the level of the scale platform to provide protection from air turbulence often prevalent in the intensive care nursery environment. The third environment was a comprehensive protective layer of thin plastic interposed between the warmer and the baby's skin, and surrounding the baby completely. The plastic blanket we used was polyethylene (saran), transparent to infrared heat and light, pliable and mouldable around the infant. We attempted to suspend the plastic over blanket rolls in such a way that the blanket did not consistently come into contact with the infants skin, potentially resulting in masceration.

We measured air velocity, insensible water loss and radiant power density as previously described. In figure 13.-11, air velocity was demonstrated to decrease incrementally from the control environment when compared to the walled chamber, and was least with the plastic blanket in place. Diminishing air turbulence is shown in polygraph recordings to the right of this figure. Insensible water loss (Figure 13.-12) paralleled diminishing convective turbulence experienced by the infants, decreasing from 3.2 (control) to 2.4 (walled) to 1.8 (blanket) ml/kg/hr. This latter number for the weight range of babies in this study is similar to that demonstrated for premature infants nurtured within humidified incubators, (Table 13.-2). Radiant power density (Figure 13.-13) was similar as long as convective air turbulence was present at the baby's skin and did not diminish appreciably until the infant was completely encased within the saran plastic blanket.

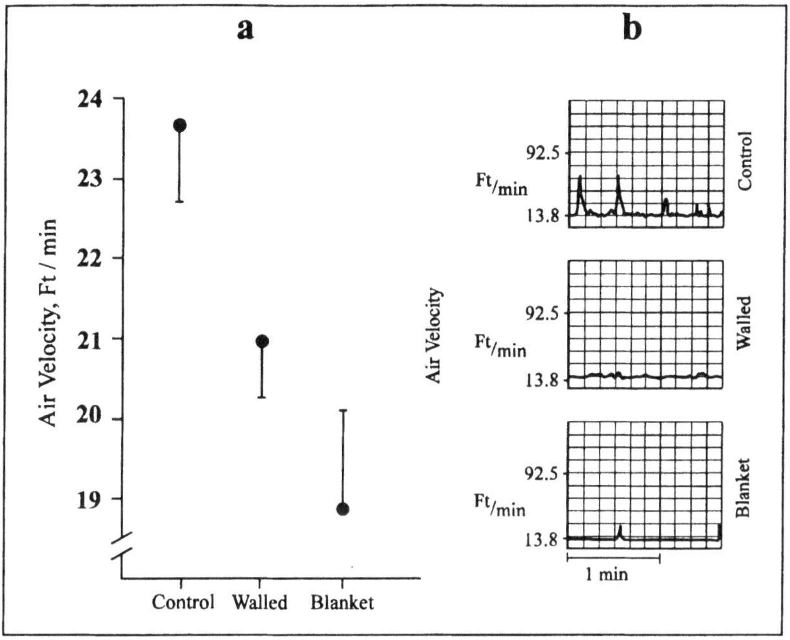

Fig. 13.-11 : Altering the radiant warmer bed configuration using either plastic walls surrounding the infant, or a thin, flexible saran plastic blanket placed loosely over the infant diminishes forced convective air turbulence experienced by babies incrementally, when compared to nursing them naked and supine under the heating element. (With permission, Mosby-Year Book Inc., Baumgart *et al*, Journal of Pediatrics, 1981; 99: 948)[25]

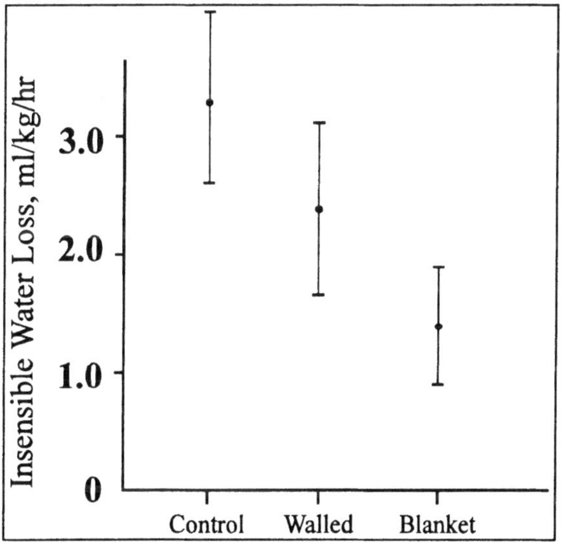

Fig. 13.-12 : Diminishing insensible water loss parallels decrements in convective turbulence with alteration of the infant's environment under the radiant heater using either plastic walls surrounding the infant, or a thin, flexible saran plastic blanket placed loosely over the infant. (With permission, Mosby-Year Book Inc., Baumgart *et al*, Journal of Pediatrics, 1981; 99: 948)[25]

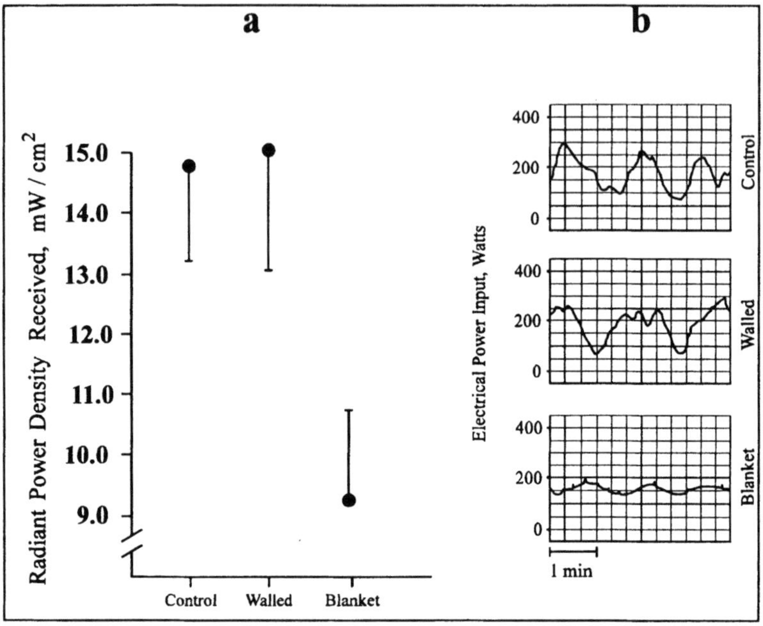

Fig. 13.-13 : Radiant energy also decreases with environmental manipulation, however, only when the saran blanket is in place is servocontrolled radiant heat delivery reduced significantly. (With permission, Mosby-Year Book Inc., Baumgart *et al*, Journal of Pediatrics, 1981; 99: 948)[25]

We concluded from this study that perhaps radiant energy does not superheat the skin, increasing insensible water loss. Rather, lower ambient air temperature and relative humidity is present near the infant's skin boundary, and with turbulence, results in a lower ambient vapor pressure. The concept of a boundary layer microenvironment (or vapor pressure gradient) comprising an envelope of protective humidity near the infant's skin, may be useful for designing optimal infant incubation strategies under radiant warmers. The more homogenous the microenvironment is near the infant's body, the less these delicate vapor and temperature gradients are disturbed.

Heat Shielding Infants under Radiant Warmers: Plastic Body Hoods.

Alternative heat shielding has been variously described by many authors. In particular the use of rigid plastic body hoods (1-3 mm in thickness) to encase infants within an incubator-like environment under open radiant warmer bed platforms have been suggested by Yeh *et al* [23], and Bell *et al* [63]. In these two studies, Yeh *et al* described a decrease in insensible water loss achievable under radiant warmers from 2.35 to 1.70 ml/kg/hr, while Bell *et al* failed to repeat this phenomenon in a group of somewhat smaller premature infants nursed under radiant warmers, observing 3.40 and 3.37 ml/kg/hr with and without a plastic body hood. Such conflicting results demonstrate clinical variation which may only be explained by more complete descriptions of environmental conditions imposed on infants nurtured in these complex situations.

We investigated the use of a non-occluded plastic hood positioned several centimeters over an infant's torso and sides as a means to prevent air turbulence near the baby, and thereby to conserve evaporative water loss [28]. We again measured radiant power density received at bed level, insensible water loss, and convective air currents near the baby within the plastic hood. Convective currents were variously increased or decreased depending on the location of the anomometer sensor. Not infrequently, convection cells were created around the open ends of the plastic hood (head and foot) in such a way that air turbulence moved through the chamber, around the baby's torso.

Radiant power density received under these devices *in vitro* is shown in figure 13.-14. Under a control condition with no heat shield interposed between the infant and the radiant heater, radiant energy increased proportionately as warmer power electrical input was increased. With the plastic hood in place, nearly 85 % of the radiant power output of the radiant warmer was absorbed and re-radiated by the plastic hood in all directions.

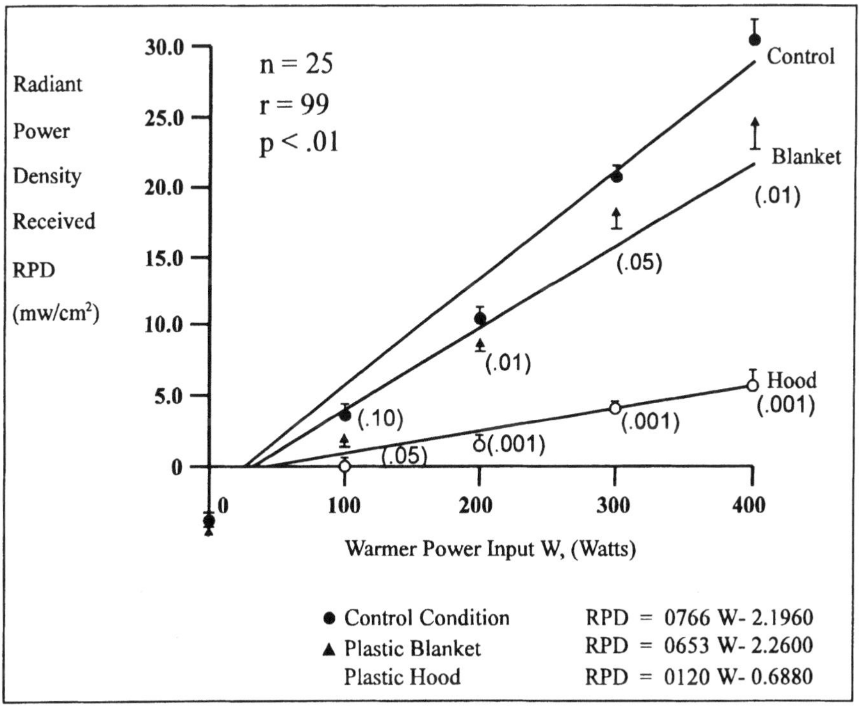

Fig. 13.-14 : Comparison of a rigid plastic body hood (several millimeters in thickness) *versus* the saran plastic blanket with respect to heat penetrance. The plastic hood absorbs a majority of the heat from the warmer, re-irradiating it in all directions at a longer wavelength. The result is a heat-sink effect rendering the servocontrol mechanism sluggish in response to changes in the infant's skin temperature. (With permission, Mosby-Year Book Inc., Baumgart *et al*, Journal of Pediatrics, 1982; 100: 787) [28]

Understandably, some of this re-irradiated energy would strike the baby's skin, however a significant portion would be re-irradiated to the cooler surrounding nursery walls. The response of the radiant warmer's servocontrol mechanism was such that infants were capable of maintaining homeothermy despite the interruption of radiant power density delivered to bed level. Shown also in this figure, the saran plastic blanket (which is much thinner, and therefore more heat transparent) interfered less with radiant power delivery, (<15 % decrement).

In vivo, (Figure 13.-15), insensible water loss was the same with both the control and hood environments, depending on convective turbulence. The servocontrolled radiant warmer power input increased significantly with use of the hood, and time delays were observed in achieving thermal equilibrium with placing and removing this device. The plastic hood comprised a heat sink obstructing the warmer's response to the infant's heat need. In contrast, the thin saran blanket significantly reduced insensible water loss, as well as the net radiant power density required to maintain infant temperature homeostasis. Radiant power density delivered at bed level was thus more homogeneously controlled.

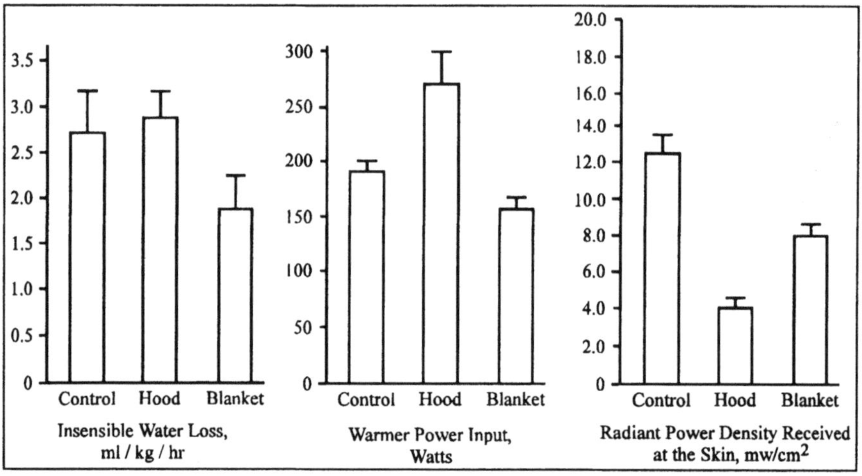

Fig. 13.-15 : Comparisons of warmer power input, heat delivered to the infant's skin, and insensible water loss in preterm babies nurtured under radiant warmers either without a plastic heat shield, or with either a rigid plastic hood versus the thin, saran plastic blanket. The plastic blanket serves as a more effective shield in preventing water loss, while significantly reducing both radiant warmer power input required, and radiant power density delivered to the baby at bed level. (With permission, Mosby-Year Book Inc., Baumgart *et al*, Journal of Pediatrics, 1982; 100: 787)[28]

Saran Plastic Blankets as Heat Shields.

Additionally, figure 13.-16 demonstrates the effect of the plastic blanket on facilitating homeothermy for infants nurtured under radiant warmers [31]. At the left, an infant is nursed naked and supine under a radiant heater. Electrical power input to the radiant warming device fluctuates between 100 and 400 watts, generating between 0 and 36-40 mW/cm^2 radiant power density at bed level. In contrast, with the saran shield in place as shown at the right, mean power input fluctuates very little around a mean of 170 watts, resulting in less radiant power density fluctuation. Less power density fluctuation implies less overdamping in the servocontroller's function, and the infant's radiant thermal experience is probably more temperate once the heat shield is in place.

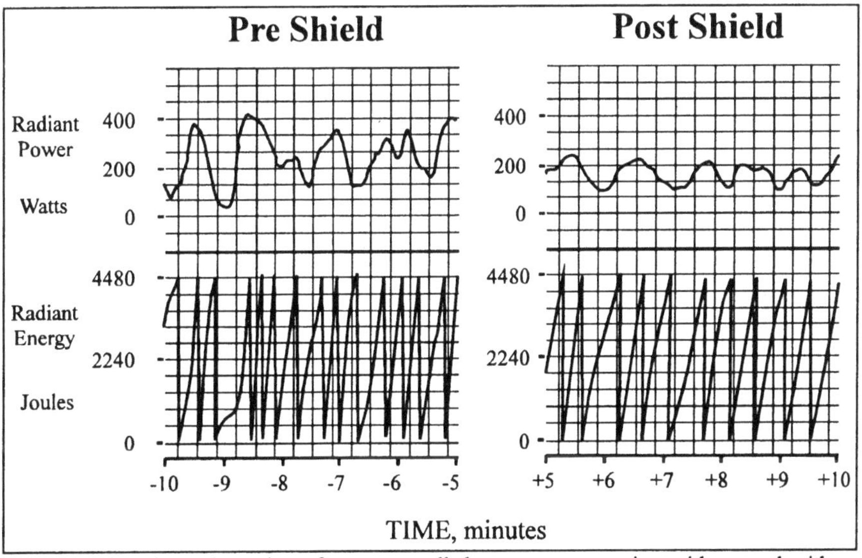

Fig. 13.-16 : Polygraph tracing of servocontrolled power consumption without and with a saran plastic blanket inplace over infants nurtured under radiant warmers. The blanket diminishes both the irradiance and the cycling peak amplitude and frequency of the controller's function. The result is a more temperate and homogenous radiant environment for infants nurtured with the blanket in place. (With permission, Williams and Wilkins, Baumgart *et al*, Critical Care Medicine, 1980; 8: 721) [31]

We further investigated the effects of the saran blanket heat shield on oxygen consumption [64]. And as shown in figure 13.-17, infants nursed naked or under the saran blanket not only experienced diminished power density requirements as shown at the right, and in insensible water loss as shown in the middle, but also experienced an 11 % decrease in oxygen consumption, presumably as a result of being nurtured under the plastic blanket. Infants in this study, however, may have been recovering from sub-optimal hypothermia, as the plastic heat shielding was used in recovering some infants from a 35.5-36.5 °C servocontrol temperature. We nevertheless conclude that the saran plastic blanket results in significantly less radiant heat need for infants to maintain homeothermy, infants incurred less evaporative water and heat losses under plastic blankets, and finally a significant conservation of oxygen consumption suggests a more temperate, homeothermic environment when nurtured under such a device.

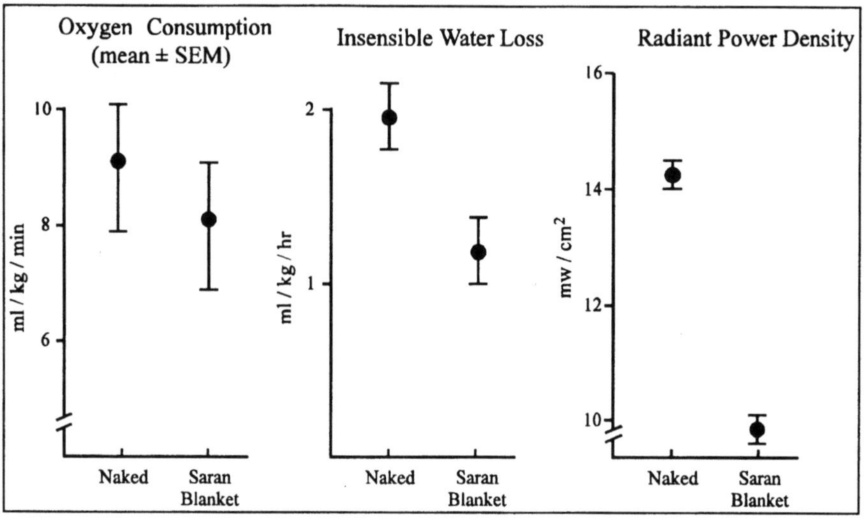

Fig. 13.-17 : Comparison of radiant power density delivered, insensible water loss and oxygen consumption in preterm babies nurtured either naked or covered by the saran plastic blanket while under servocontrolled radiant warmers. The blanket significantly reduces all three measurements consistently in all babies tested. (With permission, American Academy of Pediatrics, Baumgart, Pediatrics, 1984; 74: 1022)[64]

Non-occlusive "Artificial Skin".

Similar conservation of insensible water loss under radiant warmers has recently been demonstrated using semipermeable (non bio-occlusive) polyurethane plastic barriers adhered by polyacrylade gel directly to the infant's skin, (™Opsite or ™Tegaderm) [65]. Temperature, EKG, and oxygenation may be monitored through such artificial skin dressings, however, their effects on partitional calorimetry have not been demonstrated. These dressings reduce the magnitude of insensible water loss significantly while infants are nurtured under radiant warmers. Moreover, in a recently abstracted preliminary study by Porat *et al*, in very low birth weight infants <800 grams at birth, the amount of intravenous fluid replacement required under radiant warmers and in incubators was reduced, and the incidence of patent ductus arteriosum and mortality was significantly reduced with use of non-occlusive dressings [66]. Until these beneficial effects are confirmed, and additionally the possibility of skin trauma or infections incurred by the use of such adherent shielding are investigated, their routine application cannot be recommended.

A Thermal Neutral Environment under Radiant Warmers

Another untoward effect of radiant warmers recently cited is that metabolic rate of oxygen consumption may be increased when infants are nursed with these devices, rather than a convectively heated incubator [29]. LeBlanc *et al* in 1982 performed a meta-analysis on three studies comparing metabolic rate within convectively heated single walled incubators *versus* radiant warmers. In combining these data, LeBlanc observed that at steady state conditions, the minimal observed metabolic rate (MOMR) was increased under radiant warmers from 6.84 to 7.45 ml/kg/min. LeBlanc felt this increase of 0.6 ml/kg/min (9 %) comprised a metabolic response to thermal stress inherent with use of these devices. In practice, however, steady state conditions are rarely achieved. Thermal stress may be unavoidable in incubators as well, with repeated opening of the incubator's environment for critical care access. Bell and Rios in a study of incubator servocontrol function demonstrated low air temperature followed by over-damping of incubator air temperature control after the side panel was opened [15].

Moreover, minimal observed metabolic rate (MOMR) comprises sampling only 10-30 minutes of measured oxygen consumption while infants are observed to be at rest and apparently asleep, are 2 hours postprandial, and are nurtured at a thermal neutral temperature. Practically, fluctuations in behavior, and in environmental temperature occur continuously in the nursery which are not assessed using MOMR determinations. An estimation of daily metabolic rates based on different conditions physiologically for infants is described in table 13.-3. Basal metabolism comprises the majority of metabolic caloric generation, approximately 60 kcals/kg/day (3 watts/kg). Specific dynamic action (SDA) comprises a minimum of 10 kcals/kg/day (0.5 watts/kg).

Tab. 13.-3 : Physiologic partition of neonatal metabolic rate into estimated thermodynamic categories.

Category	Kcal/kg/day	Watts/kg
Basal metabolism	60	3.0
Specific dynamic heat	10	0.5
Activity	20	1.0
Thermal equlibration	20	1.0
Growth	10	0.5

Activity and thermal exposure contribute further to the infant's metabolic energy expenditure, and finally growth requires approximately 10 kcals/kg/day (0.5 watts/kg). Clinicians generally estimate total neonatal metabolic energy expenditure at approximately 120 kcals/kg/day.

Recently, Karl Schultz, *et al* have proposed that a more complete description of infant metabolism is derived by observing different sleep, feeding, and external activity states over 12-24 hour measurements of oxygen consumption [55, 56]. We have compromised by measuring metabolic rates of oxygen consumption in premature infants for 90-180 minute periods, *ie* during one complete feeding cycle [20, 41, 64]. Infant behavior sampling for this duration seems adequate to represent the infant's global experience of his/her environment.

We evaluated the effect of temperature at three servocontrol set points of 35.5, 36.5 and 37.5 °C in a group of 18 premature infants nurtured under radiant warmers in order to sample behavior during an approximate feeding cycle for 1-2 hours [20]. Results of continuously measured oxygen consumption in one infant are shown in the top graph of figure 13.-18, with characteristic non-steady state peaks and valleys in metabolic rate. Heart rate in the second graph is shown to increase and decrease as activity and behavior state fluctuate in parallel.

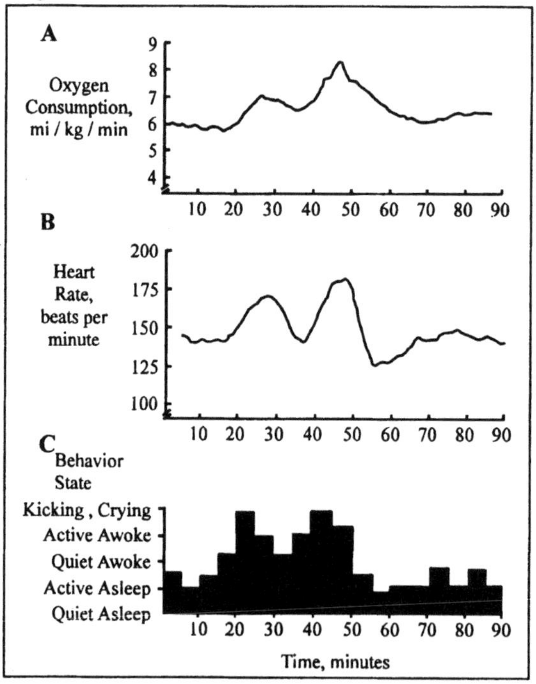

Fig. 13.-18 : Metabolic rate of oxygen consumption parallels heart rate and infant activity as expected with the performance of indirect calorimetry in a premature baby of approximately 1.0 kg weight. (With permission, American Academy of Pediatrics, Baumgart, Pediatrics, 1985; 75: 89)[41]

In figure 13.-19, for this same infant, examples of infant metabolism over comparable time periods are shown additionally at 36.5 and 37.5 °C servocontrol temperatures. Metabolic rate of oxygen consumption was significantly diminished when the radiant warmer was controlled to 36.5 °C, and was statistically (but not physiologically) significantly diminished further by a further temperature increase to 37.5 °C.

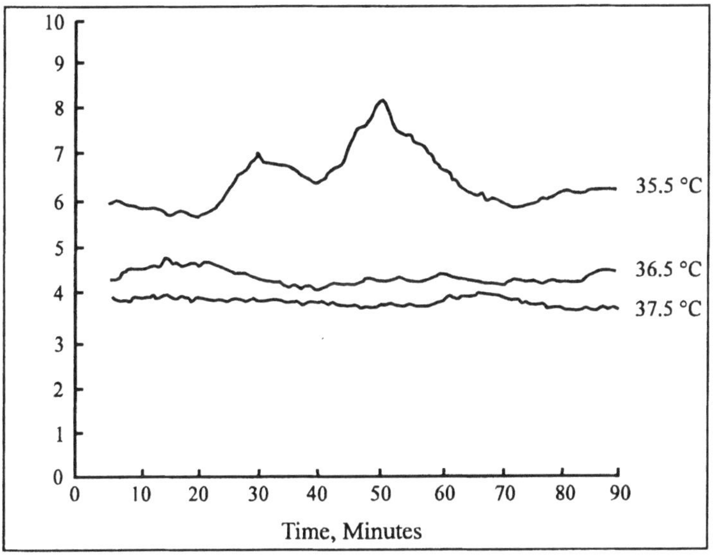

Fig. 13.-19 : The same infant's metabolic rates of oxygen consumption are compared at 35.5, 36.5 and 37.5 °C servocontrol skin temperature. Peak activity is attenuated, and basal metabolism is significantly reduced over the 90 minute duration of this study. (With permission, American Academy of Pediatrics, Malin and Baumgart, Pediatrics, 1987; 79: 47)[20]

Figure 13.-20 demonstrates the summary of all such observations for the 18 infants studied over all three temperatures tested. A statistically significant, and probably physiologically significant decrease in metabolic rate of oxygen expenditure may be achieved (generating a thermal neutral environment) for infants nurtured at an abdominal wall skin temperature of 36.5 °C. Further increases above 37.5 °C did not produce any further conservation of metabolic rate. Interestingly, the median behavior state of babies nurtured under radiant warmers was also reduced at an appropriate thermal neutral temperature, indicating that infant behavior is influenced by their thermal experience. We have observed that infants who are thermally comfortable are less active, are more likely to assume extended postures, breathe and have heart rates with lower variability, and are more likely to spend greater than 80 % of the time sleeping when nurtured in a thermal neutral fashion under radiant warmers.

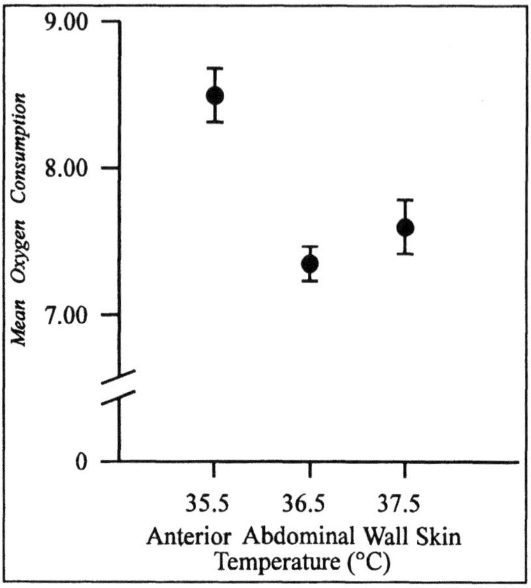

Fig. 13.-20 : Significantly lower oxygen consumption is demonstrated at 36.5 °C when compared to 35.5 °C servocontrol temperature set-point for a series of 18 infants nurtured naked and suppine under radiant warmers. No further benefit in reduced oxygen consumption is demonstrated by increasing servocontrol temperature to 37.5 °C. (With permission, American Academy of Pediatrics, Malin and Baumgart, Pediatrics, 1987; 79: 47)[20]

Finally, increase of temperature above the 36.5 °C set point resulted in significant hyperthermia in a minority of the subjects investigated. As shown in figure 13.-21, the temperature gradients between the skin, the core and the servocontrol set point diminishes as servocontrol temperature is increased towards what is considered normal. Once this gradient reverses, and the skin becomes hyperthermic, fever results. We conclude that a moderate skin temperature between 36.5 and 37 °C would therefore probably best correspond to a thermal neutral zone for infants nurtured naked and supine on an open bed platform under a radiant warmer. Clearly, there is risk for hyperthermia when infants are nurtured much above this level.

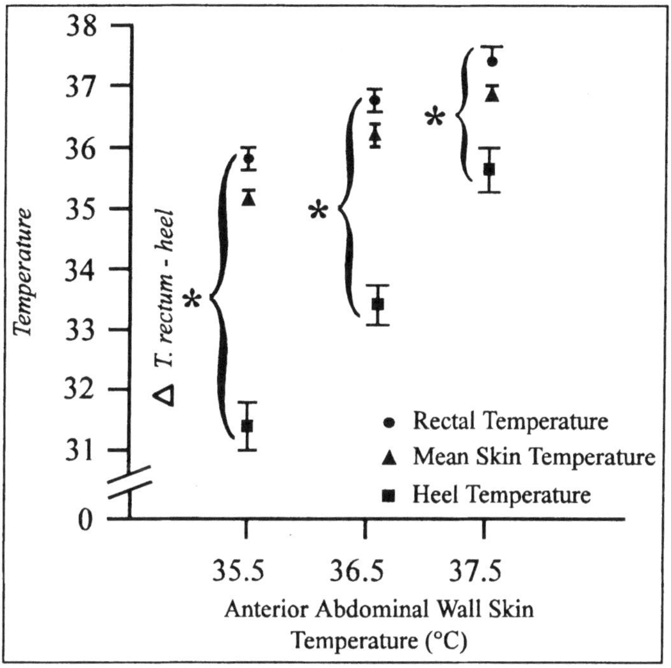

Fig. 13.-21 : Skin *versus* core temperature gradient is reduced at warmer servocontrol temperature for infants nurtured under radiant warmers. At 37.5 °C, several infants actually reversed this gradient and became hyperthermic, such that the experiment was terminated. (With permission, American Academy of Pediatrics, Malin and Baumgart, Pediatrics, 1987; 79: 47)[20]

Other Potential Advertise Effects of Infrared Radiation

Particular concern has been expressed that infrared exposure of the immature skin and eyes of premature human infants nurtured under high powered radiant heaters might produce adverse effects. Radiant warmers tend to generate the majority of their heat output in the far infrared portion of the electromagnetic spectrum between 2.0 and 20 microns. Nevertheless, a significant amount of near infrared spectral energy between 0.7 and 2.0 microns, or even visible red light between 0.5-0.7 microns may be delivered at high heater outputs. Violet and ultraviolet emissions from radiant warmers are negligible. This spectral emission range contrasts the sun or incandescent lighting which may emit ultraviolet radiation from 0.2 microns, to far infrared spectral irradiance. Infants themselves are black body irradiators, and tend to emit solely in the far infrared range of the electromagnetic spectrum between 7.0 and 20 microns.

Figure 13.-22 from Sliney and Worbarsht's text on laser energy demonstrates a biologic definition of these electromagnetic spectral wave length ranges [67]. Adverse effects are shown as photokeratitis and retinal burns in the ultraviolet spectrum, and corneal burns occurring with prolonged radiant energy exposures at high levels in the far infrared spectral regions. Cataracts may occur with exposure in the ultraviolet spectrum, and also in the infrared near and far wave length spectral ranges when prolonged, (*e.g.*, as observed in glass blowers and steel furnace workers). Anterior eye structures seem most vulnerable to infrared effects. Retinal injury, *eg* with sun gazing, is probably mediated by ultraviolet exposure which occurs concomittantly with infrared penetration to this depth in the near infrared range. Thermal skin burns are seen with virtually any radiation overdose within the electromagnetic spectrum. The reference line for tissue penetration and depth is shown at the bottom of this figure.

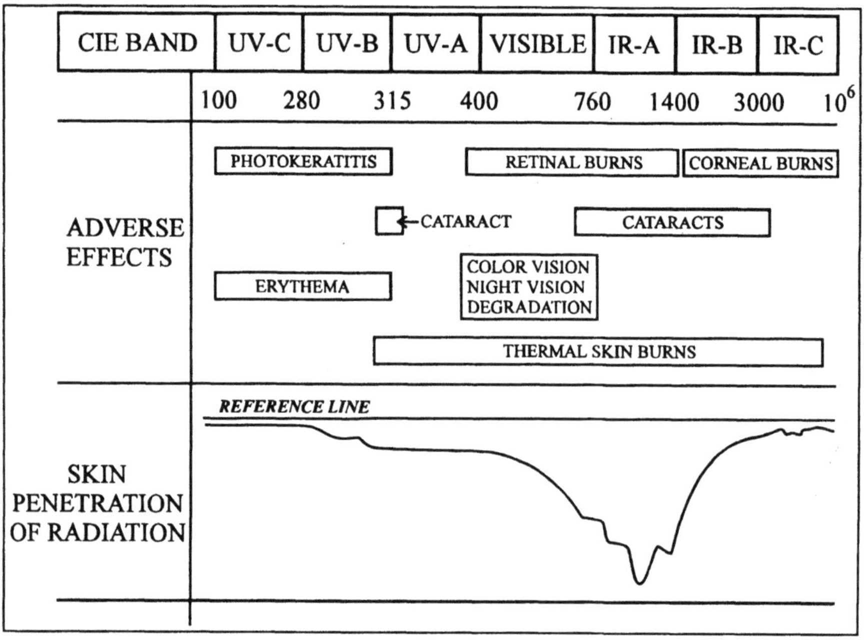

Fig. 13.-22 : Biologic absorption of electromagnetic radiation across the spectrum, including infrared heat. Potential adverse effects observed with quantitatively extreme exposures are cited. (With permission, Plenum Publishing Co., Sliney and Worbarsht, 1980; 176: 144)[67]

In a study undertaken in collaboration with the Air Shields/Vickers Corporation, we have recently published results of spectral irradiance measurements made under commercially available radiant warmers by an independent investigator [30]. Figure 13.-23 demonstrates the spectral irradiance of the Air Shields Infant Care System (Hatboro, PA). Irradiance is shown at wavelenghts from 1,000 to greater than 4,500 nanometers. Two peaks are noted in spectral infrared heat delivery. The first peak at 2,000 nanometers constitutes emission from the wire element (in the middle portion of the infrared electromagnetic spectrum). The smaller peak at the right (longer wave length spectrum beyond 3,000 nanometers) probably represents re-irradiation of heat from the quartz envelope surrounding the wire element. The irradiance levels in the parallel curves shown comprise different electrical power input levels from lowest to maximum rated power. Interestingly, the notch in the near peak probably represents environmental water vapor's absorption of energy on that particular day of testing.

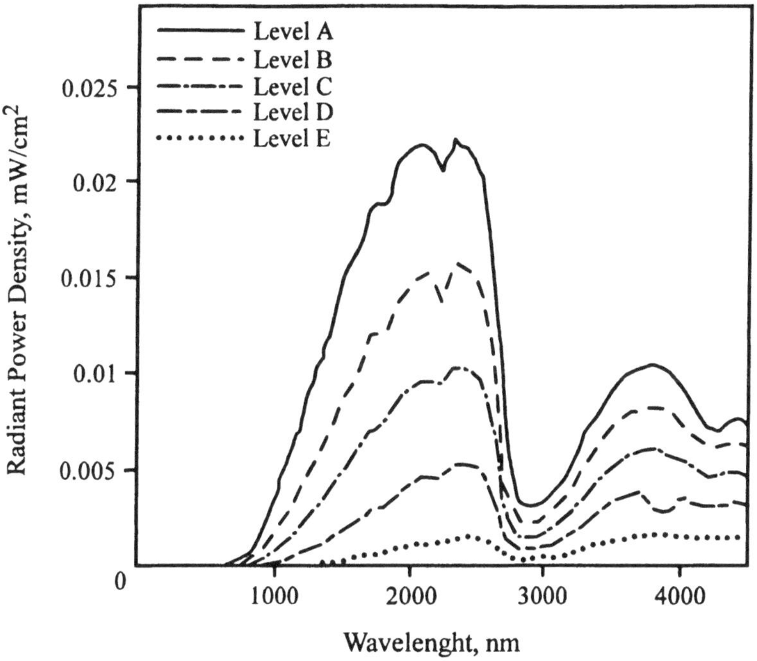

Fig. 13.-23 : Spectral irradiance measured quantitatively at bed level, and at increasing power inputs to a radiant warmer's heating element. Note two peaks of energy indicating the wire element's emission (near infrared wavelength spectrum), and the quartz containment tube's emission (far spectrum). The notch probably indicates infrared adsorption by water content of the air intervening between the warming element at bed level. (Adapted with permission, American Medical Association, Baumgart *et al*, American Journal Diseases in Children, 1993; 147: 565)[30]

Further reported in this study, a data base looking for ophthalmologic complications of prematurity (obtained during the NIH collaborative study on vitamin E prophylaxis for retinopathy of prematurity) was queried to provide 1-6 year follow-up examinations in 47 infants who were nurtured either exclusively under radiant warmers (33 babies), or exclusively within incubators (14 babies). As many variables as possible were matched to determine that these infants were comparable, however, the degree of illness for radiant warmer babies produced a bias for higher mortality in this group. None of the infants was observed to have anterior chamber eye lesions attributable to radiant energy exposure (cataracts, corneal opacities, or corneal ulcerations) in either the radiant warmer or the infant incubator group (Table 13.-4). Retinopathy of prematurity (lower grades I-II IROP classification) was more prevalent in the infants nurtured under radiant warmers, and one infant was blind (grade V) at long-term follow-up.

Tab. 13.-4 : Summary results of serial ophthalmologic examinations in infants under radiant warmers, versus incubator group, mean ± SD. (With permission, American Medical Association, Baumgart *et al*, Am J Dis Child, 1993; 147: 565) [30]

	Radiant Warmer	Convection Incubator	P<
Number infants	33	14	
Cataracts	0	0	NS
Transient vacuoles	1	2	NS
Corneal opacity	0	0	NS
Corneal ulcers	0	0	NS
Birth hemorrhages	4	0	NS
Retinopathy of prematurity:			
None	9	7	0.1
Grade I	2	2	NS
Grade II	18	5	0.1
Grade III	3	0	NS
Grade IV	0	0	NS
Detachment	1	0	NS
Age diagnosis ROP (days)	46±13	37±16	0.1
Late sequelae:			
None	19	9	NS
Motility or refractive	6	4	NS
Above and ROP	7	0	0.1
Blind with ROP	1	0	NS
Duration follow-up (months)	33	38	NS
Range follow-up	(1-75)	(1-63)	

Follow-up data in all study infants demonstrate that cataract is a relatively rare occurrence. The general population's incidence of congenital (or acquired) cataracts is from 0.4 to 0.8 % of the population. To detect a two-fold increase in this incidence, over 1,500 subjects would have to be randomized in a long-term study to answer the question whether radiant warmers contribute to anterior eye injury in a situation where no such toxicity has yet been reported. The best biologic data currently available would suggest that a total infrared spectral radiant power density delivered at full power to infants under radiant warmers should be less than 50 mW/cm^2. Most currently marketed radiant warmers deliver less than 40 mW/cm^2 at peak power. *In vivo* measurements to maintain skin temperature (and therefore core temperature) even in the very smallest low birth weight infants requires radiant power density means between 22-25 mW/cm^2. Every effort should be made to maintain maximum exposures tightly controlled within this range. Near infrared wave length exposure less than 10 mW/cm^2 is also desirable. Most radiant warmers emit far less than this amount of near infrared spectral irradiance.

Summary

Radiant energy is an efficient, continuous means for warming very premature infants under a variety of non-homeothermic, non-steady state conditions, including the neonatal intensive care unit required for the survival of such vulnerable subjects. The untoward effects of infant radiant warming are a potential for increased insensible water loss and a potential for increased infant metabolic rate required to produce thermal neutral conditions. These liabilities may be readily countered by appropriate shielding techniques, provided these shielding techniques render the infant's environment homeothermic and are left in place undisturbed. The art of infant radiant warming is in the development and application of these shielding techniques, and in the avoidance of environmental invasion when not clinically necessary. In any environment, infants are best left undisturbed when critically ill. If free from physiologic and environmental disturbances, such infants (even when on mechanical respirators) may remain comfortable and pain free. Our best intentions should be directed towards achieving these goals.

Implications for Practice

1. Until proven otherwise, the fetal thermal equilibrium should be considered desirable for survival and growth.

2. Artificial incubation following premature birth should approximate (as closely as possible) the intrauterine environment. This is probably advisable in both health and disease.

Recommendations for Future Research

1. Is not a single, fetal temperature (both surface and core) determined by maternal core temperature (amniotic fluid and placental blood flow)?

2. What is maternal core temperature, (37.2 - 38.3 °C)?

3. How can we measure and minimize heat exchange gradients during extrauterine incubation:

   skin - to - air ?

   skin - to - wall ?

   evaporation (humidity) ?

   conduction (insulation) ?

4. Is application of excess heat necessary and advisable:

   air temperature above body temperature?

   mattress temperature above body temperature?

   radiant heat above body temperature?

5. How can we test for "best" infant outcomes:

   to achieve optimal body temperature(s)?

   to achieve homeothermy?

   to minimize metabolic energy expenditure resulting from

   extrauterine exposure?

   to maximize growth?

   to maximize survival?

   to audit adverse outcomes, (e.g. hyperthermia, infection)?

References:

1. Buetow KC, Klein SW (1964) Effect of maintenance of "normal" skin temperature on survival of infants of low birth weight. Pediatrics ; 34: 163.

2. Davies PA, Davis JP (1970) Very low birth weight and subsequent head growth. Lancet ; 2: 1216.

3. Day RL, Caliguiri L, Kamenski C, *et al* (1965) Body temperature and survival of premature infants. Pediatrics ; 34: 171.

4. Jolly H, Molyneux P, Newell DJ (1962) A controlled study of the effect of temperature on premature babies. J Pediatr ; 60: 889.

5. Silverman WA, Fertig JW, Berger AP (1958) The influence of the thermal environment upon the survival of newly born premature infants. Pediatrics ; 22: 876.

6. Adamson K, Gandy GM, James LS (1965) The influence of thermal factors upon oxygen consumption of the newborn human infant. J Pediatr ; 66: 495.

7. Scopes JW, Ahmed J (1966) Minimal rates of oxygen consumption in sick and premature infants. Arch Dis Child ; 44: 407.

8. Scopes JW, Ahmed J (1966) Range of critical temperatures in sick and premature newborn babies. Arch Dis Child ; 44: 417.

9. Hey EN (1969) The relation between environmental temperatures and oxygen consumption in the newborn baby. J Physiol ; 200: 589.

10. Hey EN, Katz G (1970) The optimum thermal environment for naked babies. Arch Dis Child ; 45: 329.

11. Saur PJJ, Dane HJ, Visser HKA (1984) New standards for the neutral thermal environment of healthy very low birth weight infants during the first week of life. Arch Dis Child ; 59: 18.

12. Silverman WA, Sinclair JC, Agate FJ (1966) The oxygen cost of minor changes in heat balance of small newborn infants. Acta Paediatr Scand ; 55: 294.

13. Bell EF, Gray JC, Weinstein MR, et al (1980) The effects of thermal environment on heat balance and insensible water loss in low birth weight infants. J Pediatr ; 96: 452.

14. Bell EF, Rios GR (1983) Air versus skin temperature servocontrol of infant incubators. J Pediatr ; 103: 954.

15. Bell EF, Rios GR (1983) Performance characteristics of two double walled infant incubators. Crit Care Med ; 11: 663.

16. Agate FJ, Silverman WA (1963) The control of body temperature in the small newborn infant by low-energy infrared radiation. Pediatrics ; 31: 725.

17. Levison H, Linsao L, Swyer PR (1966) A comparison of infrared and convective heating for newborn infants. Lancet ; 2: 1346.

18. Silverman WA, Sinclair JC, Scopes JW (1966) Regulation of body temperature in pediatric surgery. J Pediatr Surg ; 1: 321.

19. Price HV, Whelpton D, McCarthy J (1971) Newborn infant heater. Lancet ; 2: 1294.

20. Malin SW, and Baumgart S (1987) Optimal thermal management for low birth weight infants nursed under high-power radiant warmers. Pediatrics ; 79:47-54.

21. Williams PR, Oh W (1974) Effects of radiant warmer on insensible water loss in newborn infants. Am J Dis Child ; 128: 511.

22. Wu PYK and Hodgeman JE (1974) Insensible water loss in preterm infants: Changes with postnatal development and non-ionizing radiant energy. Pediatrics ; 54: 704.

23. Yeh TF, Amma P, Lilien LD, et al (1979) Reduction of insensible water loss in premature infants under the radiant warmer. J Pediatr ; 94: 651.

24. Bell EF, Neidich GA, Cashore WJ, *et al* (1979) Combined effect of radiant warmer and phototherapy on insensible water loss in low birth weight infants. J Pediatr ; 94: 810.

25. Baumgart S, Engle WD, Fox WW, Polin RA (1981) Effect of heat shielding on convective and evaporative heat losses, and radiant heat transfer in the premature neonate. J Pediatr ; 99: 948-956.

26. Baumgart S, Engle WD, Fox WW, Polin RA (1981) Radiant warmer power and body size as determinants of insensible water loss in the critically ill neonate. Pediatr Res ; 15: 1495-1499.

27. Baumgart S, Langman CB, Sosulski R, Fox WW, Polin RA (1982) Fluid, electrolyte and glucose maintenance in the very low birthweight infant. Clin Pediatr ; 21(4): 199-206.

28. Baumgart S, Fox WW, Polin RA (1982) Physiologic implications of two different heat shields for infants under radiant warmers. J.Pediatr ; 100:787-790.

29. LeBlanc MH (1982) Relative efficacy of an incubator and an open warmer in producing thermoneutrality for the small premature infant. Pediatrics ; 69: 439.

30. Baumgart S, Knauth A, Casey FX, Quinn GE (1993) Infrared eye injury not due to radiant warmer use in premature neonates. Am J Dis Child ; 147: 565-569.

31. Baumgart S, Engle WD, Langman CB, Fox WW, Polin RA (1980) Monitoring radiant power in the critically ill newborn under a radiant warmer. Crit Care Med ; 8: 721-724.

32. Engle WD, Baumgart S, Langman CB, Fox WW, Polin RA (1980) Radiant energy monitoring and insensible water loss under phototherapy: A new method to study mechanics in fluid balance. Pediatr Res ; 14: 596A.

33. Baumgart S, Lavanier J, Schwartz JB, Hall SB (1984) Attenuation of warming and cooling cycles by shielding thermistor probes in infants nursed under radiant warmers. Adv Ther; 1:19-25.

34. Baumgart S (1987) Current concepts and clinical strategies for managing low birthweight infants under radiant warmers.In, Medical Instrumentation ; 21: 23-28.

35. Baumgart S (1988) Thermal Regulation. In, Guidelines for Perinatal Care, 2nd Ed. American Academy of Pediatrics and American College of Obstetrics and Gynecology. Evanston, Illinois and Washington, DC.

36. Baumgart S (1991) Thermal regulation in the premature neonate. In, Schaffer & Avery's Diseases of the Newborn, 6th Ed., Taeusch HW, Ballard RA, Avery ME (eds.), Philadelphia, WB Saunders Company, pp 255-259.

37. Baumgart S (1992) Thermal regulation of the fetus and newborn. In, Neonatal and Fetal Medicine - Physiology and Pathophysiology, 1st Ed. Polin RA and Fox WW (eds), Orlando, WB Saunders Co, pp 477-526.

38. Baumgart S (1992) Incubation of the Human Newborn Infant.In, Issues in Clinical Neonatology. Pommerance J, Richardson CJ (eds), Norwalk, CT, Appleton & Lange, pp 139-150.

39. Baumgart S (1995) Perinatal and neonatal thermoregulation.In, Intensive Care of the Fetus and Neonate. Spitzer AR (ed), Chicago, Mosby Year Book Inc.

40. Baumgart S (1984) The physiology and physics of heat exchange in the premature newborn infant nursed under a radiant warmer.J Calif Perinat Soc ; IV(2):15-18.

41. Baumgart S (1985) Partitioning of heat losses and gains in premature newborn infants under radiant warmers. Pediatrics ; 75:89-99.

42. Baumgart S (1990) Radiant heat loss *v* radiant heat gain in premature neonates under radiant warmers. Biology of the Neonate ; 57:10-20.

43. Baumgart S (1984) Physics of heat exchange in the neonate.In, Medical Technology for the Neonate. AAP and AAMI Proceedings, Technology Assessment Report #9, pp 1-3, Arlington, VA, 1984.

44. Wheldon AC, Rutter N (1982) The heat balance of small babies nursed in incubators and under radiant warmers. Early Human Development ; 6: 131.

45. Wheldon AE (1982) Energy balance in the newborn baby: Use of a manikin to estimate radiant and convective heat loss. Phys Med Biol ; 27: 285.

46. Bell EF, Rios GR (1983) A double walled incubator alters the partition of body heat loss of premature infants. Pediatr Res ; 17: 135.

47. Okken A, Blijhan C, Franz W, *et al* (1982) Effects of forced convection of heated air on insensible water loss and heat loss in preterm infants in incubators. J Pediatr ; 101: 108.

48. Colin J, Houdas Y (1967) Experimental determination of coefficient of heat exchanges by convection of human body. J Appl Physiol ; 22: 31.

49. Lewis HE, Foster AR, Cox RN, *et al* (1969) Aerodynamics of the human microenvironment. Lancet ; 1: 7609.

50. Isenschmid Die Bestimmung der Wasserbilanz am Krankenbett. Med Klin Berlin, 1918; 14: 1128.

51. Sosulski R, Polin RA, Baumgart S (1983) Respiratory water loss and heat balance in intubated infants receiving humidified air. J Pediatr ; 103:307-310.

52. Doyle LW, Sinclair JC (1981) Thermal effect on a Potter baby scale. Pediatr Res ; 15: 685.

53. Darnall RA Jr (1981) Insensible weight loss measurements in newborn infants: Possible over estimation with the Potter baby scale. J Pediatr ; 99:794.

54. Lusk G The Elements of the Science of Nutrition, ed 4. Philadelphia, WB Saunders, 1928, pp 64-68.

55. Schulze K, Kairan R, Stefanski M, *et al* (1981) Spontaneous variability in minute ventilation oxygen consumption and heart rate of low birth weight infants. J Appl Physiol ; 50: 1098.

56. Schulze K, Kairan R, Stefanski M, *et al* (1981) Continuous measurement of minute ventilation and gaseous metabolism of newborn infants. Pediatr Res ; 15: 1111.

57. Haycock GB, Schwartz GJ, Wisotsky DH (1978) Geometric method for measuring body surface areas: a height-weight formula validated in infants, children and adults. J Pediatr ; 93: 62.

58. Fanaroff AA, Wald M, Gruber HS, *et al* (1972) Insensible water loss in low birthweight infants. Pediatrics ; 50: 236.

59. Williams PR, Oh W (1974) Effects of radiant warmer on insensible Water loss in newborn infants. Am J Dis Child ; 128: 511.

60. Engle WD, Baumgart S, Schwartz JG, Fox WW, Polin RA (1981) Combined effect of radiant warmer power and phototherapy on insensible water loss in the critically ill neonate. Am J Dis Child ; 135(6): 516-520.

61. Engle WD, Baumgart S, Fox WW, Polin RA (1982) Effect of increased radiant warmer power output on state of hydration in the critically ill neonate. Crit Care Med ; 10: 673-676.

62. Baumgart S (1982) Radiant energy and insensible water loss in the premature newborn infant nursed under a radiant warmer. Clin Perinatol, 9(3):483-503, W. B. Saunders.

63. Bell EF, Weinstein MR, Oh W (1980) Heat balance in premature infants: Comparative effects of convectively heated incubator and radiant warmer, with and without plastic heat shield. J.Pediatr ; 96: 460.

64. Baumgart S (1984) Reduction of oxygen consumption, insensible water loss and radiant heat demand with use of a plastic blanket for low birthweight infants under radiant warmers. Pediatrics ; 74:1022-1028.

65. Knauth A, Gordin M, McNelis W, Baumgart S (1989) A semipermeable polyurethane membrane as an artificial skin in the premature neonate. Pediatrics ; 83:945-950.

66. Porat R, Brodsky N (1993) Effect of Tegaderm use on outcome of extremely low birth weight (ELBW) infants. Pediatr Res ; 33: 231A.

67. Sliney D, Worbarsht M (1980) Safety with Lasers and Other Optical Sources. Plenum Press, New York, NY ; 176: 144.

Chapter 14

Treatment of Sick Newborns with Heated Mattress

RAGNAR TUNELL

In 1970 Hey et al [1] showed that the use of a heated mattress in a cot significantly (22 %) reduced the oxygen consumption in preterm babies. The mattress used by Hey was of insulating material with low heat transfer capacity. The thermal control of this device was however inexact. Beds with insulating mattresses but with heating elements or electric bulbs below the mattress have been widely used both in Europe (Wärmebettchen) and in developing countries. Many of these devices, however, do not fulfil modern requirements for safety.

The Heated Waterfilled Mattress:

During the last 10 years we have examined a commercially available heated water filled mattress (HWM) produced by KanMed Sweden [2].

A polyvinylchloride mattress is filled with 8 litres of water and heated by a thermo controlled heating plate (24 v AC and 50 W). The temperature of the water is regulated between 35 and 38 °C. The baby is kept on the mattress and covered with a blanket. If the baby is below 1 kg of body weight the head is covered with a hat. The control unit has a temperature display giving the actual water temperature. The upper surface temperature deviates about 0.8 °C from the displayed temperature.

The temperature is regulated by thermistors in the heating element and set to a maximum temperature of 41 °C. The system is calibrated at an ambient temperature of 24 °C. There is an influence on the water temperature of 0,08 °C for each °C of deviation from the ambient temperature. This deviation will be obvious from the temperature display but it would be preferable to regulate the temperature of the water rather then that of the heating plate.

The low electrical power (50 w) makes the heating slow (4,4 °C per hour) so the mattress is kept heated when not in use. The high heat storage capacity of the water means that cooling of the mattress, if the electrical current is interrupted, takes several hours.

The Thermal Influence from Room Temperature

The thermal conditions in a cot with HWM is affected by the room temperature in a similar way to an air heated incubator (Fig 14.-1). Unless the room temperature is 25 °C or more there is an increased heat loss from babies with both types of heating devices.

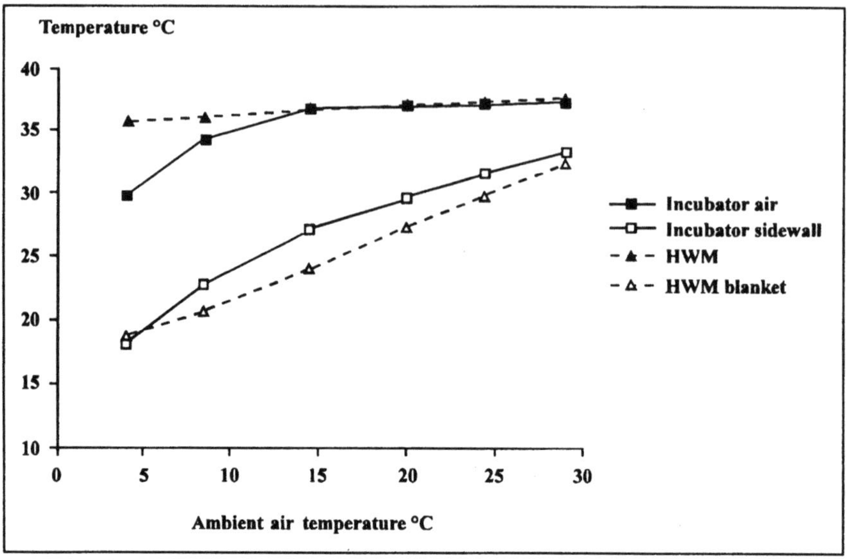

Fig. 14.-1 : Comparison of temperatures in an HWM and in a single-walled air-heated incubator in a climate chamber at ambient temperatures between 5 and 30 °C. From ref. [2] with permission from : Bio Med. Eng. & Computing.

In order to study the thermal effect a *thermal manikin* was used [3]. The manikin had the body proportions of a one kg baby. The non evaporative heat loss was measured from 8 regions of the surface which was set to a temperature 36.5 °C. The power to the different zones of the manikin was transmitted and measured from a DC power supply. All measurements were done in a climate chamber at two different ambient temperatures: 25 °C and 15 °C.

Results: When the climate chamber temperature was set at *25 °C* : the total heat loss was 20W/m^2 from the manikin on the HWM with the use of a thick blanket (The Insulating Factor of the blanket was 0.42 m^2 °C /W) The mattress temperature was 36.5 °C. In a closed incubator at 36 °C air temp the total heat loss was 13 W /m^2 and increased to 20 w/m^2 at 34.5 °C as air temperature.

When the climate chamber temperature was set at 15 °C : the manikin on the HWM had a hat and was covered by the thick blanket. The non-evaporative heat loss was 35 W/m^2 when the mattress temperature was 37.5 °C The same heat loss (35 W/m^2)was obtained in the incubator with an air temperature of 34 °C. With an air temperature of 36 °C the heat loss was 20 W/m^2. Opening of one porthole at this ambient temperature increased the heat loss to 33 W/m^2 and opening the main aperture to 82 W/ m^2.

Conclusion: Irrespective of whether the infant is treated on the HWM or in a single-walled incubator the room temperature has to be kept at least at 25 °C. There are good reasons to assume that this is also true for the so called Kangaroo-method.

Clinical Studies on Preterm Babies

An investigation was performed at Huddinge Hospital on babies after a stabilisation period in an incubator. Seventeen babies were enrolled in the "study group" treated on the HWM and 17 in the "control group" nursed in air-heated single walled incubators. The birth weight varied between 1 and 2 kg and the study period lasted 3 weeks [4].

The results showed that there was no difference between the groups in basic metabolic rate measured as oxygen consumption, rectal temperature, mean skin temperature and daily weight gain. The temperature of the water in the mattress was close to 36.5 °C. The air temperature of the incubator was a mean 33.5 °C on starting and this decreased by 0.9 °C during the study.

Conclusions: : Babies not needing intensive care or to be observed naked may, from a thermal point of view, be well treated in a cot with a heated water mattress.

In a second study, also performed at Huddinge Hospital, 12 preterm babies were alternately cared for in an air heated incubator or in a bed with HWM. During routine care foot, abdomen and axillary temperatures were measured repeatedly. Metabolic rate was measured during both types of care as basic oxygen consumption. Continuous monitoring of heart rate was performed during the study [5].

The results showed that, during cleaning and weighing of the babies, the foot temperature dropped in both types of care but in the cot care the drop was significantly shorter that in incubator care - Figure 14.-2. No difference was observed in the other skin temperature measurements. The metabolic rate, determined as basic oxygen consumption, was significantly lower during cot nursing. During undisturbed periods the cot nursed children had significantly lower heart rate than the incubator cared children.

Conclusions Cot nursing provides an environment which is less disturbing for the babies than incubator care, with less thermal stress on routine procedures and more rest when undisturbed.

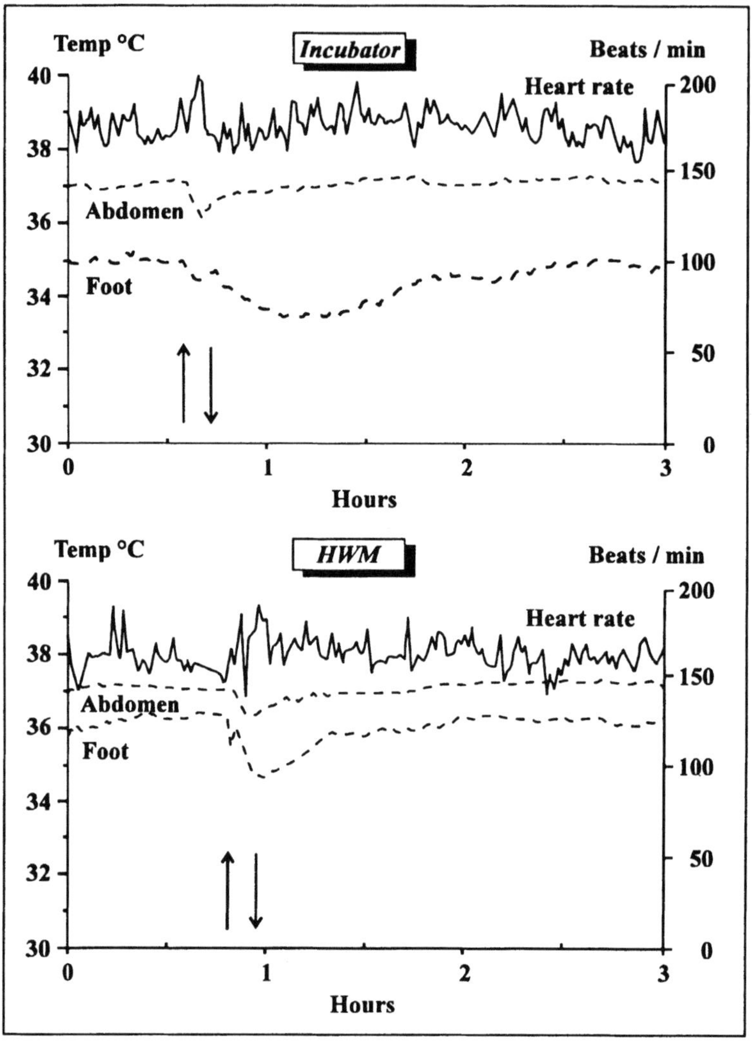

Fig. 14.-2 : Three hour recordings of heart rates and temperatures (Foot and abdomen) in the same infant during routine care in incubator and on the HWM. Arrows indicate start and end of intervention with nappy change. From ref. [5] with permission from Acta Paediatrica.

Rewarming hypothermic babies was investigated in Turkey (Capa University Clinic in Instanbul) [6]. Sixty infants weighing between 1 and 2 kg were randomly selected to treatment in an air heated incubator or in a cot with the HWM. The mean body temperature on admission was 33 °C and 90 % of the babies had a rectal temperature between 30 °C and 36 °C. The axillary temperatures were recorded 3 times daily and in some infants every hour - Figure 14.-3.

Results :In the 28 babies treated on the HWM the axillary temperature was normal on the day of admission but in the incubator group normalisation took 3 days. Contributing to this were frequent electrical failure, badly functioning incubators and frequent opening of portholes. The mortality was also lower in the "mattress group" (21 %) compared to the "incubator group" (34 %).

Conclusion. In a neonatal unit where there is a lack of qualified personal and well functioning incubators the use of cot nursing with heated mattress might be preferred.

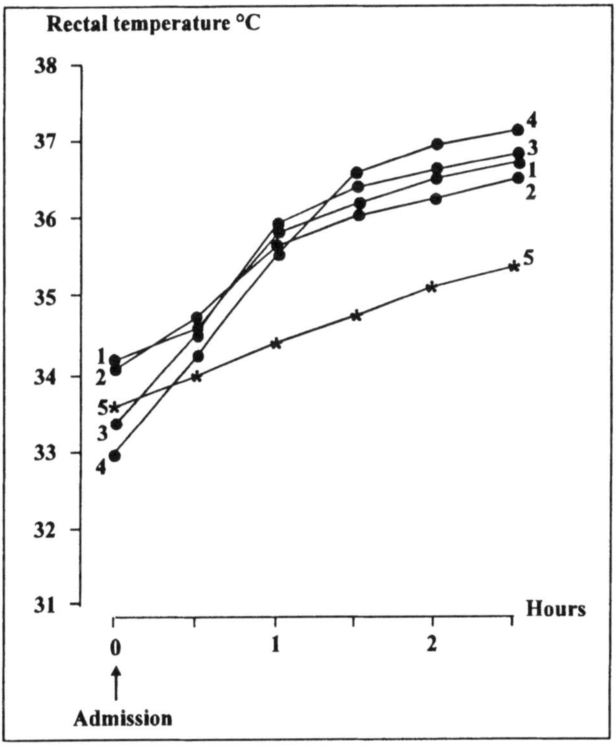

Fig. 14.-3 : Rewarming of five hypothermic infants . Cases no. 1-4 were treated on the HWM and case no. 5 in an airheated incubator. From ref. [7] with permission from Arch. Dis. Childh.

Mothers perception of the care of the infants was investigated [7]. Ten Swedish mothers and 23 Turkish mothers were interviewed about their attitude and feelings towards their new-born babies. Repeated interviews were performed by a psychologist during the stay in the hospital.

Results: Both the Swedish and the Turkish mothers expressed lack of contact with their babies in the incubator group and increased anxiety. On the other hand when treated in the cot on the HWM they had easy access to the babies and felt that the babies thrived well. At the time of the investigation the use of skin to skin contact had not been introduced.

Conclusion: Improved contact between mother and child is important during incubator care. This can be achieved by using cot-nursing with a HWM or by using skin to skin contact provided the room temperature is at least 25 °C.

The HWM as Supplemental Heat Supply under Overhead Radiant Warmer.

In 1983 Thomas and Topper [8] studied 8 preterm infants (BW 710-1040 g) during intensive care. Five received ventilatory support. The infants were studied as there own controls. They were covered with a thin plastic shield and had a skin servo control for the overhead heater (set point :36 °C). The water mattress was kept at 37 °C. Different skin temperatures, axillary temperature and the wattage output from the radiant heater were recorded with and without water mattress.

Results: During the 2 hour test period with the mattress the wattage from the radiant heater was significantly reduced by a mean of 21 % with also much less variability in wattage. The foot temperature was also higher during the period with the water mattress.

Conclusion: Conductive heat transfer has a significant influence on the thermal environment for infants cared for under a radiant warmer.

The Use of HWM as Supplement to Single Walled Incubator.

In 1984 Marks et al [9] reported the results from 10 healthy infants with gestational age 28-33 weeks who were randomised to incubator care with or without supplement of heat from a waterfilled heated mattress with 35 °C of water temperature. Oesophageal and skin temperature measurements as well as oxygen consumption was recorded.

Results: There was an increased heat storage during the period with the water mattress and a significant increase in oesophageal temperature from 36,5 to 37,2 °C and mean skin temperature from 35.3 to 36.3 °C. No difference in oxygen consumption was observed.

In the neonatal unit at Huddinge Hospital we have routinely used the HWM with the water temperature set at 37 °C inside incubators when treating babies below one kg of birth weight. This has made it possible to use incubators where the air temperature is maximised to 35-36 °C.

Conclusion: The use of a heating mattress inside an air heated incubator or under a radiant heater stabilises the body temperature of the very low birth weight baby.

Complications with the Use of a HWM.

The quality of the thermocontrol of the mattress has to be very exact. The thermal safety range is narrow at 35 - 38 °C. Above this range there is risk of overheating and burns and below of cooling. It is also very important that the temperature is distributed evenly and not influenced by part of the heating element covered by insulating material. Water has the effect of distributing the heat evenly and thus makes the mattress more safe.

The quality of the mattress has to be very good. The recommendation from the firm KanMed is to replace the mattress every third year to avoid risks of leakage. The cot must have holes in the bottom to let water out if a leak should occur. Inside the incubators a leak might create electrical damage. At present it is not accepted by international safety standards to use water mattresses inside air heated incubators.

New devices, using non-rigid plastics, have been developed for heating mattresses in radiant heated open beds and for surgical use on operating tables. It might well also be introduced for incubators.

Heated Mattress during Transport

In 1976 Nielsen et al [10] used a mattress producing heat via exothermic crystallisation reaction initiated by injection of a small amount of catalyst. The mattress increased the thermal condition of a transport incubator and improved rewarming of hypothermic babies.

At Huddinge Hospital we have used the warm water mattress separated from the heating pad and placed it inside a Dräger transport incubator for the transportation of very sick low birth weight babies needing assisted ventilation during transport (30 min transport time in road ambulance). With this method hypothermia during transport has been avoided completely.

Conclusion: To our knowledge contact heat transfer during transportation is little used but deserves more attention. If the ambient temperature is low and it is necessary to open portholes then cooling is unavoidable unless extra heat is provided from a heated mattress.

The Use of Skin to Skin Contact

Skin to skin contact was first practised in Colombia by Ray and Martinez [11] for very low birth weight babies (below 1500g BW). In 1989 Whitelaw et al [12] investigated 14 very low birth weight babies at a mean postnatal age of 35 days. The infants were naked and in close skin to skin contact with their mothers, covered with a blanket and a hat. In three of the smallest babies careful temperature recordings took place. The room temperature was 26-29 °C.

Results: In the three babies the back skin temperature was unchanged or increased during the intervention.

Implication for Practice

1. Contact heat transferred either from a water mattress or skin to skin provides an effective method to stabilise the body temperature in preterm babies provided.

- That the quality of the mattress is good with efficient heat transfer and even distribution of warmth.

- That the thermoregulation of the contact surface is exact and not below 35 °C or above 38 °C.

- That the baby is covered with a thick blanket and, in where the body weight is below 1000 g, a hat is used.

- That the room temperature is at least 25 °C.

2. Under a radiant heater or inside stationary- and transport - incubators the introduction of an effective warm mattress results in increased thermal stability for the very low birth weight babies. Further development is needed however to get a safe, effective and well thermocontrolled contact mattresses for this use.

References:

1. Hey EN, O'Connell B (1970) Oxygen consumption and heat balance in the cot-nursed baby. Arch Dis Child 45: 335-43

2. Sarman I, Tunell R Ribbe T (1993) Thermally controlled water-filled mattress for warming preterm infants :a physical assessment. Med.& Biol. Eng. & Comput., 31: 639-643

3. Sarman I, Bolin D, Holmér I, Tunell R (1992) Assessment of thermal conditions in neonatal care: Use of a manikin of premature baby size. Am J of Perinatol 9: 239-246

4. Sarman I, Tunell R (1989) Providing warmth for preterm babies by a heated, water filled mattress. Arch Dis Cild 64: 29-33

5. Sarman I (1992) Thermal responses and heart rates of low-birth weight premature babies during daily care on a heated water-filled mattress. Acta Paediatr 81: 15-20

6. Sarman I, Can G, Tunell R (1989) Rewarming preterm infants on a heated, water filled mattress. Arch Dis Child 64: 687-692

7. Sarman I, Tunell R, Västberg I, Can G, Toparlak D (1993) Mothers perception of their preterm infants treated in an incubator or on a heated water-filled mattress: a pilot study. Acta Paediatr 82 930-933

8. Topper W, Stewart T (1984) Thermal support for the very-low-birth-weight infant:Role of supplemental conductive heat. J Pediatr 105: 810-814

9. Marks K, Calder P, Nardis E, Ultman J (1984) A warming mattress for premature infants.Ped.Res 18: 333A

10. Nielsen H, Jung A, Atherton S (1976) Evaluation of the porta-mattress as a source of heat for neonatal transport. Pediatrics 58 550-504

11. Whitelaw A, Sleath K (1985) Myth of the marsupial mother:hoem care of very low birth weight babies. Lancet I : 1206-1208

12. Whitelaw A, Heisterkamp G, Sleath K, Acolet D, Richars M (1988) Skin to skin contact for very low birthweight infants and their mothers. 63: 1377-1381

Chapter 15

Aspects of Hygiene

HEIKE MARTINY

Neonatal Infections

Different studies report infection rates between 5.2 % and 30.4 % for hospital-acquired infections in newborn intensive care units [21, 22, 29, 30, 31, 66]. All neonates should be considered as immune-compromised, and in particular the preterm, those with congenital malformations, or infants infected before - transplacental, during or after birth. When treated in pediatric or neonatal intensive care units there is a significant increase in numbers of infections for a variety of reasons.

The immature skin and mucous membranes as well as the stump of the umbilical cord are all excellent portals of entry for micro-organisms [40]. The lack of normal flora, defects in host resistance, the immature immune system, low birth weight, and parenteral feeding [19, 21, 36, 44, 52] all contribute to an increased risk of infection.

Another important factor is the treatment administered to the neonate. Even during routine care of healthy neonates there is a risk of transmission of micro-organisms. In the absence of normal flora is lacking colonization with micro-organisms will result more frequently to infection than in other patients. If treatment necessitates medical equipment such as ventilators, humidifiers, scalp electrodes, or incubators the risk of infection is increased. In addition, it has to be assumed that the risk of a transmission of micro-organisms increases proportionally not only to the time spent in hospital but also to the number of different people involved in administering treatment.

The underlying mechanism of transmission is similar for micro-organisms which are known to be pathogenic and those which are only potentially pathogenic and there is little to be gained by separating bacteria, fungi or viruses. The different modes of transmission will therefore be discussed in general, to highlight the possible sources of infection and to stress the infection risks from which the neonate must be protected. In addition it must not be forgotten that the neonate is not only in danger of being infected but can also, as a carrier of micro-organisms, transmit infection to other babies and the staff.

Sterility

Sterilisation of equipment and instruments must be effective and the technique used regularly checked and validated. The material must remain sterile until it is used and this may require the use of packaging which protects from recontamination. The requirements of the users must be decided in advance of purchase of sterilizing equipment and care taken to make sure the manufacturer can meet the necessary specification.

If sterile material is not completely used at one time, (e.g. medication in multiple-dose containers), contamination must be avoided and the date and time when the container was opened recorded.

Disinfection

The purpose of disinfection is to reduce the number of pathogens to a point where infection can no longer occur. Since neonates are particularly susceptible to infection, disinfective measures should be of a high standard. The number of micro-organisms surviving after disinfection is dependent on the size of the initial colony and a standardized disinfection procedure which always uses the same concentration and duration of exposure will always reduce the number of organisms by the same amount.

As well as being affected by colony size, disinfection will be less successful in the presence of residual blood or protein. A further point to be taken into consideration when choosing a disinfection procedure is the target group of micro-organisms. It may be necessary, for example, to ensure that the procedure destroys the hepatitis B viruses or mycobacteria.

The use of automatic washer-disinfectors for (chemical-) thermal treatment allows the use of higher temperatures and lower concentrations of disinfectant. Thermal treatment can be used for instruments (including some made of plastic, such as endotracheal tubes), utensils (feeding bottles, teats, milk pumps) or laundry (sheets, protective clothing, cleaning cloths). It is advisable to make sure such materials are thermally stable before they are purchased.

Surfaces of changing tables, scales, and instruments can only be disinfected manually and it is recommended that the cleaning is combined with the disinfection process. If routine disinfection is not done it is essential that the micro-organisms removed from the surfaces cleaned are not simply being dispersed further and that they cannot multiply on the cleaning utensils.
Dispensing with disinfecting agents will rarely help cut costs and most of the agents are easily degradable active components of disinfectants such as formaldehyde or peracetic acid. For these reasons dispensing with routine disinfection cannot be recommended.

Implication for Practice

1. Disinfection of Hands and Skin

Without any doubt hands are the principal factor in the transmission of micro-organisms, both bacteria and viruses [6, 7, 24, 35, 38, 45, 50, 51, 65]. Hand washing alone will reduce by a factor of 10^3 the transient skin flora but the organisms will not be destroyed and may be further transmitted from sinks and protective clothing. If the unit's hygiene policy considers that hand washing alone is sufficient, it must be sure that the remaining transient skin flora is safe for the neonate, and that the hands of the person administering treatment are not colonized with other pathogenic or potentially pathogenic micro-organisms from conditions such as dermatitis. Care must be taken to ensure there is no transmission of organisms from soap and towels.

If disinfectants are used these must be products with a proven effectiveness. In Europe a substance used for hand disinfection must be capable of reducing in vitro a selection of vegetative bacteria, in a practice-oriented exposure time (generally 30 secs), by a factor of 10^5. Thus successful disinfection requires that the concentration and the exposure time are checked and complied with. In addition, it is vital that the active component of the disinfectant will actually be effective against the micro-organism for example, alcohols are ineffective against viruses without a membrane, such as rotavirus. Contaminated disinfectants suggest that the active components are either in the wrong concentration or are ineffective.

The wearing of gloves protects the wearer from direct contact with any foreign flora, so long as the gloves are not defective, but it does not prevent transmission of micro-organisms. If the gloves rather than the hands are subjected to disinfection it is important to verify the resistance of the glove material to the disinfectant. In order to prevent transmission of micro-organisms disinfection of hands or gloves is needed prior to and following each contact with the neonate.

When balancing hand disinfection against hand washing, it should be taken into account that the available hand disinfectants tend to produce fewer skin irritations than hand washing as they contain moisturising substances. The skin is not soaked in water as long as it is during washing which also means that less skin care is needed.
Disinfectants containing iodine must not be used on neonates as they can cause transitory hypothyroidism. Except for compounds containing silver (silver acetate [Credé] or silver nitrate), disinfectants containing metals compounds should not be used. For prophylactic disinfection prior to performing invasive procedures such as punctures, injections, infusions or transfusions the use of alcohol-containing preparations is recommended. When caring for central lines or puncture electrode sites chlorhexidine should

be used. Whether the use of a disinfectant containing octenidine dihydrochloride, an active agent with a broad range of action, is in accordance with national regulations must be verified in each individual case.

2. Invasive Measures

Without doubt any invasive procedure provides a direct portal of entry for micro-organisms or allows infection to develop through contamination. It is beyond any doubt that all medical instruments introduced into sterile areas must also be sterile. This applies to suction, intravenous and urinary bladder catheters, infusion sets and electrodes.

If an invasive measure is essential the site where it is carried out may be important in determining the risk involved to the neonate. For example in bladder catheterization studies have shown that suprapubic bladder puncture involves a lower risk of infection than urethral catheter drainage [57]. Sites of access for a central catheters differ in their infection risks. Studies have shown that colonization of jugular, subclavian, umbilical and femoral sites differ considerably in the numbers and species of micro-organisms present [10, 28, 41]. Some of these differences have proved significant and a recent study strongly recommends the subclavian vein as the access site for a central catheter [3].

Preparation of all infusions and other formulations for parenteral use involves a risk of contamination. Any micro-organism introduced into such a solution will readily multiply. The risk of infection then depends on the susceptibility of the recipient as well as the number of micro-organisms present as well as on their pathogenicity. The rate of multiplication is greatly influenced by the nutrient concentration, the oxygen content and the temperature of the solution.

3. Infusions

As already mentioned the subclavian vein should be selected as the access sites for central catheters in neonates. Strict aseptic techniques are essential for insertion. The effectiveness of routine changing of catheters, recommended by some authorities as an infection-preventing measure, has not been proven in independent studies [59]. In Germany, according to a court judgement concerning the preparation of infusions, medication may only be added up to a maximum of 1 hr before administration. The preparation of additives on a clean bench can only be considered an infection-preventing measure when great care is taken to keep the working surface unencumbered and disinfected as well as regularly checking the effectiveness of the bench filters. The use of (endotoxin) in-line filters does not offer any real security from infection as the amount of endotoxin retained depends not only on the amount present before the filter but also on the type of infusion solution used. Also any addition of suspensions, emulsions, cell-containing solutions or medication in low concentrations can only be added after the filter [68].

4. Intubation

Endotracheal or nasotracheal intubation is carried out under strictly aseptic conditions using sterile material (tube, laryngoscope). For suctioning sterile material is also required, i.e. disposable sterile catheters. Since a tube will be colonized with micro-organisms while it is in place it must be remembered that these patients are not only at high risk of infection, but also represent a source of infection as they are colonized by micro-organisms and are also more likely to have an infection of the airways [33, 60].

5. Gastric tubes

In neonates only sterile gastric tubes should be used since any alteration of gastric pH may result in a massive colonization of the stomach contents making the baby in turn a source of infection. If sterile disposable material is not available, adequate disinfection of the tubes and other accessories is essential.

6. Feeding

Contaminated baby food may lead to infection in the neonate [47]. For example, Pseudomonas aeruginosa may provoke an epidemic of pneumonia, sepsis, osteomyelitis as well as meningitis and encephalitis. For these micro-organisms, the sources of contamination may be the personnel (hands), infected equipment, containers, cleaning utensils and possibly the drinking water.

Tests with commercial baby food that was artificially contaminated revealed it to be an excellent culture medium for micro-organisms [46]. Depending on the species of organism, the storage time and temperature, colonization can reach dangerously high figures. It must always be remembered that cells of Enterobacteriaceae, such as E. Coli, take no longer than 30 mins to double under optimum conditions. It is however not only commercial baby food that is a potential danger. Expressed breast milk may also put the baby at risk if it is not used immediately as it is an excellent medium for micro-organisms during tansportation and storage [4, 13, 31, 49].

7. Materials

As a rule surfaces should only consist of materials which can be wipe-disinfected, that is to say materials that have smooth surfaces. Spraying with disinfectants releases aerosols and cannot be approved. It is also important to consider that it is more difficult to wipe, and thus disinfect, an untreated wooden surface than a smooth surface made of metal, glass, or synthetic material. Conclusive evidence that synthetic materials can never compare to metal from a hygienic point of view has certainly not yet been found and this question requires further investigation.

When medical equipment is being designed some thought should be given to the question of cleaning and disinfection. Inaccessible and possibly humid dead space, gaps, grooves and corners should be avoided. Buttons and knobs on the operating panels of devices must be shaped in such a way, and must be made of such a material, that they can be wiped with a disinfectant.
For manual disinfecting and cleaning, as well as for thermal disinfection in automatic washer-disinfectors, not only is the surface condition of the material (e.g., the grain size of metals) of importance, but also the compatibility of the material with the both the disinfectant and the method used. If, for example, a disinfectant containing formaldehyde is used the material to be treated must be resistant to the disinfectant and this should be checked with the manufacturer. To cite an example of incompatibility from neonatology, the hoods of incubators may be damaged by treatment with surface disinfectants containing alcohol. If thermal disinfection, as the preferable method of disinfection is chosen, the temperature resistance of the material must obviously be checked prior to use, or even better before purchase.

8. Incubators

When neonates need to be placed in incubators colonization with normal flora will be delayed due to the reduced contact with their mothers and other objects in their environment [3, 25, 63]. Instead they will be colonized with different, hospital-specific flora [15, 62].
There is no conclusive evidence of a relation between the microbial colonization of the neonate and contamination of the incubator (Ayliffe). It is certain, though, that there is always a risk of transmitting micro-organisms during handling if the incubator has not been cleaned correctly (e.g., no rigorous cleaning and disinfecting before using it for a new neonate) and it is contaminated with micro-organisms foreign to the neonate's own flora [26, 55]. This risk is increased if the micro-organisms are able to multiply on the surface of the incubator, as a result of humidity or organic soiling (faeces, liquids, etc.) [48, 56].

In incubators where a water reservoir is used for humidifying the circulating air this container is generally contaminated with micro-organisms [14, 30, 37, 48, 56, 64, 67]. There is concern that these organisms could be transmitted to the neonate but as yet there is no convincing evidence that this is a major risk factor. When the same organisms are found in the humidifier and the neonate it seems likely that they in fact passed, via skin abrasions, from the baby to the water rather than the other way round [8, 53, 70]. Whether or not infection can occur from the water it is important that the humidifier is uncontaminated since transmission of micro-organisms can always occur from hands or other objects during handling. Colonies numbering up to 10^{10}/ml have been reported in the humidifier water [11, 12, 37, 56] and the quality used - sterile drinking water or distilled water - appears to have a bearing on microbial growth. To some extent residual chlorine in the drinking water seems to have a continued disinfectant activity in the incubator [70].

The temperature in the reservoir also influences the number of micro-organisms present and many will not multiply at temperatures above 40 °C [70]. The addition of chemicals can also reduce the number of micro-organisms but the addition of disinfectants to the humidifier water can only be done if the substance used will not pass into the air or if it is proven to be toxicologically safe for neonates. A safe method of humidification from the point of view of hygiene is the vaporization of sterile water. Even if the humidifier water has been contaminated and microbial growth has occurred, all micro-organisms will be destroyed in the vaporization process.

When an incubator is in use as much of the interior as possible should be wiped daily with a sterile cloth (pad, diaper) and sterile distilled water. Disinfectants should not be used routinely. Changing the humidifier water daily is only of use if the design allows the reservoir to be completely emptied. Incubators should be changed weekly[48, 56]. The cleaning and disinfecting of incubators must be of a very high standard. Thermal disinfection is the most desirable method but this is not yet possible with currently available incubators. A device for steam disinfecting incubators which was developed by Dräger did not find favour with the users [39, 43]. Therefore the only possible method available at the present time is chemical disinfection by wiping the incubator. Cleaning and disinfecting the water reservoir can be a problem as calcium deposits can prevent adequate disinfection and microbial growth can start again immediately the reservoir is refilled (see Materials above).

An incubator must be stored after cleaning and disinfection with its water reservoir completely dry. If solutions containing formaldehyde are used for wipe-disinfecting incubators, they must be rewiped with a sterile cloth (pad, diaper) and sterile distilled water following the required duration of exposure. The incubator must be then be aired for 24 hours at its operating temperature. If rewiping and/or airing are not done, it must be ensured that no residual disinfectant can pass from the wiped surfaces into the air circulating in the incubator during operation.

9. Humidification of Inspired Gases

Infections of the airways may be caused not only by aspiration or transmission of micro-organisms via the hands of personnel, but also by the spread from equipment used for inhalation therapy [23, 54].

Of the different methods for humidifying breathing gases a distinction must be made between humidifiers and nebulizers. Simple nonheated humidifiers function by the pass-over or blow-by system or they work by a bubble diffusion principle or jet system. If saturated, warmed air is needed the water in the reservoir must be heated and for this most are of the cascade or wick humidifier type.

Nebulizers produce aerosols with water droplet size of between 0.5 μm and 15 μm. From 2 μm upward the droplets inhaled via nose and mouth can pass into the bronchioles [1]. Nebulizers are operated at different temperatures up to 55 °C. Ultrasonic nebulizers use high-frequency vibrations to produce the aerosol. In contrast to the bubble diffusion system, pass-over or blow-by humidification does not seem to lead to a dispersion of micro-organisms from the humidifier water (see Incubators above).

When judging and evaluating the different systems, various criteria must be taken into account. The most important point is whether the design of the device permits safe cleaning and disinfection. When sterilization is not possible owing to the design, thermal disinfection in an automatic washer-disinfectors is recommended. If the device needs to be taken apart for sterilization and disinfection then there is a risk of recontamination during reassembly of the parts [69]. Sterile water in commercially available disposable containers seems to inhibit microbial growth and should always be used in humidifiers. It is also recommended that only small amounts of water are used at a time. If medication is necessary it must be added under aseptic conditions. When the water in a humidifier is being heated, the rise in temperature results in a decline in microbial growth, but does not completely preclude growth as there are differences in temperature in different parts of the humidifier [61].

The comparison of the different systems must not be based solely on the extent to which the humidifier water is contaminated since this does not say anything about the actual risk as the example of pass-over humidification shows. It is important to determine the absolute number of micro-organisms present in the respirable aerosol [2].

10. Thermometers

Thermometers, especially rectal, are inevitably contaminated with micro-organisms after use [58]. Sheaths protect the thermometers from contamination but they are not really necessary if a disinfection by immersion is carried out correctly. Even if patients have personal thermometers they should be disinfected after each use. Cross transmission of infections due to thermometers is known on paediatric wards [16, 32]. If thermometers made of glass are replaced by electronic measuring methods, it must be verified beforehand that these can be adequately disinfected. It is especially important that in the case of thermometers used for taking rectal temperature, an immersion disinfectant is employed which is effective against viruses, including the hepatitis B virus.

11. Stethoscopes

No study has yet shown that an infection can be traced to a contaminated stethoscope [17, 20, 42]. However because of the particular susceptibility of neonates to infection it is advisable to wipe the stethoscope with an alcohol-based disinfectant between patients, or to use a personal stethoscope for each patient.

12. Electrodes

In order to avoid transmission of micro-organisms, electrodes must be wiped with an alcohol-based disinfectant between patients [9, 40]. Contamination of electrodes may also be due to non-sterile sodium chloride solutions used for moistening.

References:

1. Andersen A (1958) New sampler for the collection, sizing, and enumeration of viable airborne-particles. U.S. Army Chemical Corps Proving Ground, Dugway. Utah 76, 471-484

2. Anschütz P-C (1994) Vergleichende Untersuchungen eines Düsen- und Ultraschall-Verneblers - Bakterien im Nebelausstoß nach experimenteller Kontamination. Inaugural-Dissertation der Freien Universität Berlin, Berlin

3. Bertone SA, Fisher MC, Mortensen JE (1994) Quantitative skin cultures at potential catheter sites in neonates. Infect Control Hosp Epidemiol 15, 315-318

4. Blenkharn JI (1989) Infection risks from electrically operating breast pumps. J Hosp Infect 13, 27-31

5. Borderon JC, Gold F, Laugier J (1981) Enterobacteria of the neonate. Normal colonization and antibiotic-induced selection. Biol Neonate 39, 1-7

6. Burke JP, Ingall D, Klein JO, Gezon HM, Finland M (1971) Proteus mirabilis infections in a hospital nursery traced to a human carrier. N Engl J Med 284, 115-121

7. Casewell M, Phillips I (1977) Hands as a route of transmission for Klebsiella species. Br Med J 2, 1315-1317

8. Clark RP (1975) Hygiene of babies' incubators. Lancet I, 696-697

9. Cefai C, Elliott TSJ (1988) Re-usable ECG electrodes - a vehicle for cross-infections? J Hosp Infect 12, 65-67

10. Cronin WA, Germanson TP, Donowitz LG (1990) Intravascular catheter colonisation and related bloodstream infection in critically ill neonates. Infect Control Hosp Epidemiol 11, 301-308

11. Davenport SE (1992) Frequency of hand washing by registered nurses caring for infants on radiant warmers and in incubators. Neonatal Network 11, 21-25

12. Donowitz LG (1987) Hand washing technique in a pediatric intensive care unit. Am J Dis Child 6, 683-685

13. Donowitz LG, Marsik FJ, Fisher KA, Wenzel RP (1981) Contaminated breast milk: a source of Klebsiella bacteremia in a newborn intensive care unit. Rev Infect Dis 3, 716-720

14. Edmondson EB, Reinarz JA, Pierce AK, Sanford JP (1966) Nebulization equipment: a potential source of gram-negative pneumonias. Am J Dis Child 111, 357-360

15. Eriksson M, Melèn B, Myrbäck KE, Windblad B, Zetterström R (1982) Bacterial colonization of newborn infants in neonatal intensive care unit. Acta Paediatr Scand 71, 779-783

16. French GL (1981) Salmonella cross-infection associated with contamination of rectal thermometers. J Hosp Infect 2, 389-390

17. Garner TK, Rimland D (1982) Stethoscopes and infection. JAMA 248, 310

18. Gaynes RP, Martone WJ, Culver DH et al (1991) Comparison of rates of nosocomial infections in neonatal intensive care units in the United States. Am J Med 91, 192-196

19. Gerdes JS, Yoder MC, Douglas SD et al (1983) Decreased plasma fibronectin in neonatal sepsis. Pediatrics 72, 877-881

20. Gerken A, Cavanagh S, Winner HI (1972) Infection hazard from stethoscopes in hospital. Lancet 1, 1214-1215

21. Goldmann DA, Durbin WA, Freeman J (1981) Nosocomial infections in a neonatal intensive care unit. J Infect Dis 144, 449-459

22. Goldmann DA, Leclair J, Macone A (1978) Bacterial colonization of neonates admitted to an intensive care environment. J Pediatr 93, 288-293

23. Goularte TA, Craven DE (1987) Bacterial colonization of cascade humidifier reservoirs after 24 and 48 hours of continuous mechanical ventilation. Infect Control 8, 200-203

24. Hambræus A, Lagerquist-Widh A, Zettersten U, Engberg S, Sedin G, Sjöberg L (1991) Spread of Klebsiella in a neonatal ward. Scand J Infect Dis 23, 189-194

25. Handrick W, Goltzsch M, Braun W, Sprencker FB (1987) Untersuchungen zur bakteriellen Besiedlung gesunder sowie kranker und/oder untergewichtiger Neugeborener. 2. Bakterielle Besiedlung hospitalisierter kranker und/oder untergewichtiger Neugeborener. Pädiatr Grenzgeb 26, 193-199

26. Handrick W, Spencker F-B, Hückel D, Bergmann L, Boehm G, Rother T (1984) Einfluß von Hygiene- und Antibiotikaregime auf die bakterielle Flora bei stationär behandelten Früh- und Neugeborenen. Z gesamte Hyg 30, 8-13

27. Harpin VA, Rutter N (1983) Barrier properties of the newborn infant's skin. J Pediatr 102, 419-425

28. Hartman GE, Shochat SJ (1987) Management of septic complications associated with silastic catheters in childhood malignancy. Pediatr Infect DisJ 6, 1042-1047

29. Hemming VQ, Overall JC, Britt MR (1976) Nosocomial infections in a newborn intensive care unit: results of forty-one months of surveillance. N Engl J Med 294, 1310-1316

30. Hoffman MA, Finberg L (1955) Pseudomonas-infect in infants associated with high-humidity environments. J Pediatr 46, 626-630

31. Hunter PR (1991) Application of hazard analysis critical control point (HACCP) to the handling of expressed breast milk on a neonatal unit. J Hosp Infect 17, 139-146

32. Im SWK, Chow K, Chau PY (1981) Rectal thermometers mediated cross-infection with Salmonella wandsworth in a paediatric ward. J Hosp Infect 2, 171-174 9

33. Inglis TJJ, Millar MR, Jones JG, Robinson DA (1989) Tracheal tube biofilm as a source of bacteria colonisation of the lung. J Clin Microbiol 27, 2014-2018

34. Jolley AE (1993) The value of surveillance cultures on neonatal intensive care units. J Hosp Infect 25, 153-159

35. Kayyali MZ, Nicholson DP, Smith IM (1972) A Klebsiella outbreak in a paediatric nursery: emergency action and preventive surveillance. Clin Pediatr 11, 422-426

36. Klein RB, Fischer TJ, Gard SE, Biberstein M, Rich KC, Stiehm ER (1977) Decreased mononuclear and polymorphonuclear chemotaxis in human newborns, infants, and young children. Paediatrics 60, 467-472

37. Kiosz D, Stoffregen C, Simon C, Stockhausen B v (1984) Über den Luftkeimgehalt auf der Intensivstation einer Kinderklinik. Mon schr Kinderheilkd 132, 274-277

38. Krajden S, Middleton PJ (1983) Enterovirus infection in the neonate. Clin Pediatr 22, 87

39. Lessky E, Martiny H, Koller W, Rüden H (1989) Evaluierung eines Systems zur thermischen Desinfektion von Säuglings-Inkubatoren. Hyg Med 14, 53-58

40. Lockey E, Parker J, Casewell MW (1973) Contamination of ECG electrode pads with Klebsiella and Pseudomonas species. Br Med J 2, 400-401

41. Maki DG, Ringer M (1897) Evaluation of dressing regimes for prevention of infection with peripheral intravenous catheters. JAMA 258, 2396-2402

42. Mangi RJ, Andriole VT (1972) Contaminated stethoscopes: a potential source of nosocomial infections. Yale J Biol Med 45, 600-604

43. Martiny H, Rüden H (1987) Zur thermischen Desinfektion von Inkubatoren: Untersuchungen über ein thermisches Desinfektionsverfahren. Bundesgesundhbl 30, 129-134

44. Miller ME (1969) Phagocytosis in the newborn infant: Humoral and cellular factors. J Pediatr 74, 255-259

45. Modlin JF (1980) Fatal echovirus 11 disease in premature neonates. Pediatrics 66, 775-780

46. Miorini T, Möse RJ, Mascher F (1988) Wachstumsverhalten nosokomialer Infektionserreger in Säuglingsnahrung. Hyg Med 13, 417-419

47. Miorini T, Sixl W, Brosch R (1988) Hospitalismus in einer neonatologischen Intensivstation. Hyg Med 13, 219-222

48. Moffet HL, Allan D, Williams T (1967) Survival and dissemination of bacteria in nebulizers and incubators. Am J Dis Child 114, 13-20

49. Moloney AC, Quoraishi AH, Parry P, Hall V (1987) A bacteriological examination of breast pumps. J Hosp Infect 9, 169-174

50. Mortimer EA, Lipsitz PJ, Wolonsky E, Gonzaga AJ, Rammelkamp CH (1962) Transmission of staphylococci between newborns: importance of the hands of personnel. Am J Dis Child 104, 289-295

51. Nyström B (1994) Impact of handwashing on mortality in intensive care: examination of the evidence. Infect Control Hosp Epidemiol 15, 435-436

52. Park BH, Holmes B, Good RA (1970) Metabolic activities in leucocytes in newborn infants. J Pediatr 76, 237-241

53. Rance CP, Roy TE, Donohue WL, Sepp A, Elder R, Finlayson M (1961) An epidemic of septicemia with meningitis and hemorrhagic encephalitis in premature infants. Am J Dis Child 101, 321-325

54. Reinarz JA, Pierce AK, Mays BB, Sanford JP (1965) The potential role of inhalation therapy equipment in nosocomial pulmonary infection. J Clin Invest 44, 831-839

55. Roncoroni AJ, Casewell MW, Phillips I (1980) The disinfection of clinically contaminated Matburn suction pumps and baby incubators in an 'Aseptor' formalin cabinet. J Hosp Infect 1,251-259

56. Sever JL (1959) Possible role of humidifying equipment in spread of infections from newborn nursery. J Pediatr 24, 50-53

57. Shaffer D, Benotti PN, Bothe A Jr, Jenkins RL, Blackburn GL (1987) A prospective, randomised trial of abdominal wound drainage in gastric bypass surgery. Ann Surg 206, 134-137

58. Smith, L, Prince HN, Johnson E (1981) Bacteriologic studies on electronic hospital thermometers. Infect Control 2, 315-316

59. Snydman DR, Donnelley-Reidy M, Perry LK, Martin WJ (1987) Intravenous tubing containing burettes can be safely changed at 72 hour intervals. Infect Control 8, 113-116

60. Sottile FD, Marrie TJ, Prough DS, et al (1986) Nosocomial pulmonary infection: possible etiologic significance of bacterial adhesion to endotracheal tubes. Crit Care Med 14, 265-270

61. Spaepen MS, Bodman HA, Kundsin RB, Berryman JR, Fencl V (1975) Microorganisms in heated nebulizers. Health Lab Sci 12, 316-320

62. Spencker FB, Braun W, Goltzsch M, Handrick W (1987) Untersuchungen zur bakteriellen Besiedlung gesunder sowie kranker und/oder untergewichtiger Neugeborener. 3. Vergleich der Besiedlungsflora gesunder mit derjeniger kranker und/oder untergewichtiger Neugeborener. Pädiatr Grenzgeb 26, 201-213

63. Spencker FB, Goltzsch M, Handrick W, Braun W (1987) Untersuchungen zur bakteriellen Besiedlung gesunder sowie kranker und/oder untergewichtiger Neugeborener. 1. Bakterielle Besiedlung gesunder Neugeborener. Pädiatr Grenzgeb 26, 185-192

64. Svalander PC (1991) Myco-contamination of incubators used for human pre-embryo culture. Fertil Steril 56, 1207-1208

65. Svennningsen NW, Békássy N, Christensen P, Kamme C (1984) Nosocomial Klebsiella pneumoniae infection: clinical and hygienic measures in a neonatal intensive care unit. Scand J Infect Dis 16, 29-35

66. Townsend TR, Wenzel RP (1981) Nosocomial bloodstream infections in a newborn intensive care unit: a case-matched control study of morbidity, mortality and risk. Am J Epidemiol 114, 73-80

67. Veríssimo A, Vesey G, Rocha GM, Marrão G, Colbourne J, Dennis PJ, Costa MS da (1990) A hot water supply as a source of Legionella pneumophila in incubators of a neonatology unit. J Hosp Infect 15, 255-263

68. Weist K, Krieger J, Rüden H (1994) Rückhaltevermögen von E. coli-Endotoxinen durch Inline-Infusionsfilter im praxisnahen Versuch. In: Fortschritte in der Hygiene und Mikrobiologie (Hrsgb.: Christiansen B, Gundermann K-O, Höller C, Rautenberg P, Ullmann U) p. 218, Pechstein Verlag, Dobersdorf

69. Wendt C, Weist K, Rüden H (1993) Case study of contamination of breathing gas humidifiers with Acinetobacter baumannii. Hyg Med 18, 99-106

70. Ziegler P (1987) Zur Problematik der mikrobiellen Kontamination von Säuglingsinkubatoren. Dissertation der Friedrich-Schiller-Universität Jena, Jena

Extreme Thermal Conditions

Chapter 16

Prevention of Neonatal Hypothermia in the Delivery Room.

KARL BAUER / HANS VERSMOLD

Introduction

The fundamental studies on the thermal condition after delivery were done 25 to 30 years ago [1, 2, 3, 4]. Today the interest in this area of neonatal care needs review because :

1. Many healthy neonates are placed in skin-to-skin contact with their mothers immediately after delivery [5] to promote mother-infant bonding. This practice was uncommon 30 years ago and its impact on body temperature needs to be evaluated.
2. We are dealing with a population of very low birth weight infants unknown at the time of the initial studies. In clinical practice a considerable number of these neonates become hypothermic in the delivery room. In a study in Berlin, 66 out of 123 (54 %) neonates below 28 weeks gestation born in 1991 and 1992 had a body temperature <36.5 °C with 15 % below 35 °C or admission to the neonatal intensive care unit [6].
3. In the developing world hypothermia is much more common. Of all infants born in the maternity hospital in Kathmandu 85 % had temperatures of <36 °C at two hours of life [7].

Prevention of Heat Loss in Healthy Term Neonates

In utero the fetus lives in a warm environment at 37 °C. Heat is a byproduct of metabolism and this has to be dissipated but this still results in a body temperature in the fetus that is 0.5 °C higher than that of the mother. After delivery there is a substantial drop in environmental temperature resulting in a heat loss as high as 80 cal/kgmin [4] in a naked term baby at a room temperature of 25 °C. The neonate reacts to this heat loss by .increasing heat production, which is about twice as high at 25 °C than in a thermoneutral environment [3]. Despite this heat loss still exceeds heat production by about 20 cal/kgmin and this results in a 2 °C drop in body temperature if a term neonate is dried but then left naked at an environmental temperature of 25°C [4, 8] (Figure 15.-1). Immediate postnatal heat loss is even greater if the baby is not dried [4] or after a delivery by caesarean section [9].

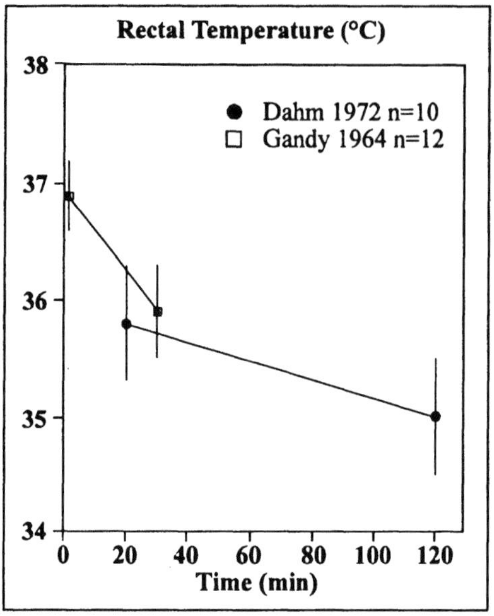

Fig. 16.-1 : Drop in rectal temperature in healthy term neonates who were dried and then left naked at a room temperature of 25 °C. [4, 8]

The thermoneutral temperature for a naked term neonate is between 32 °C [10] and 35 °C [3]. It is unrealistic to expect such a temperature in a delivery room and therefore a number of interventions to reduce heat loss in term neonates at a room temperature of 25 °C have been evaluated (Table 15.-1). For healthy neonates drying and wrapping in a warm blanket is sufficient, a radiant heater is not necessary [4].

Tab. 16.-1 : Decrease in rectal temperature in the first 30 minutes of life in term infants with various methods to reduce heat loss

Method	Decrease in rectal temp./30 min
Naked, not dried*	2.7
Naked, dried	1.0
Dried, wrapped in blanket	0.5
Dried, under radiant heater	0.4
Dried, skin to skin+	0.4

*: from Dahm et al. Pediatrics 1972

+: from Christensson Acta Paediatr 1992

Recently immediate skin contact between mother and neonate after delivery has been recommended to promote mother-infant bonding. Christensson [7] has compared body temperatures in healthy term neonates either cared for skin-to-skin or in a cot. Infants in the experimental (skin-to-skin) group were dried, placed prone on mother's abdomen and covered with a blanket. Infants in the control group were dried, swaddled in a cotton cloth and put in a cot. Both groups were at a room temperature of about 26 °C. Skin-to-skin contact with the mother resulted in the same body temperatures as wrapping in a blanket. Skin-to-skin care might even provide sufficient warmth for very low birth weight infants. Own preliminary studies in which we measured oxygen consumption during skin-to-skin care for the first time, even showed even a slight increase in body temperature without an increase in oxygen consumption during skin-to-skin care compared to baseline (Figure 16.-2).

In conclusion, for term infants, drying and wrapping in a warm blanket or skin-to-skin contact with the mother after drying are both equally effective measures to prevent heat loss.

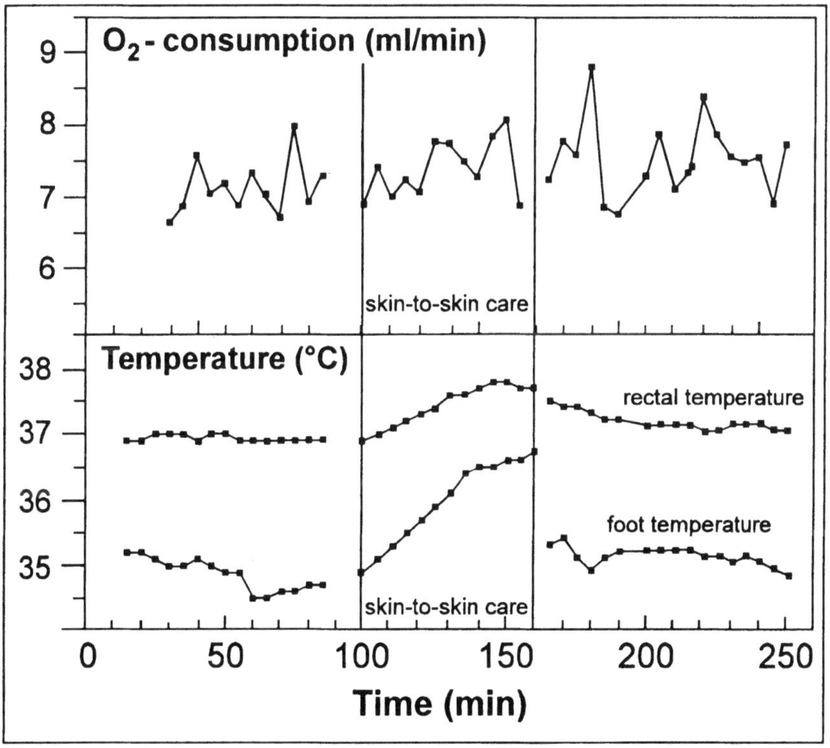

Fig. 16.-2 : O_2-consumption, rectal and foot temperature in a 4 day old 1200g preterm infant.

Prevention of Heat Loss in the Asphyxiated Infant

I wish to focus on two aspects of the relationship between heat loss and asphyxia: Does asphyxia impair the thermogenic response to cold and second, and can hypothermia be used to protect against hypoxic brain damage?

Hypoxia impairs the thermogenic response to cold. In newborn kittens the thermogenic increase in oxygen uptake could be completely blocked by reducing the oxygen content in the inspiratory air to 10 %, though their basal metabolic rate remained normal [11]. A reduced increase in oxygen uptake was also observed in asphyxiated human newborns at an environmental temperature of 25 °C [8]. Nonshivering thermogenesis is more vulnerable to hypoxia than shivering thermogenesis.

The idea of using hypothermia as a potential therapy to prevent hypoxic-ischaemic brain damage in asphyxiated neonates was stimulated by sensational case reports of survival without brain damage after prolonged accidental submersion in cold water [12]. In neonatal rats decreased hypoxic brain injury was demonstrated when the rats were exposed to hypoxia at a body temperature of 31 °C [13].

In human infants deep hypothermia is used to repair cardiovascular malformations in circulatory arrest [14] and there are reports of its use for the treatment of asphyxiated human neonates [15, 16, 17]. Hypothermia was induced by immersion in tap water (10-15 °C) for 5 minutes. Body temperatures between 30 and 32 °C were achieved [17].

Despite the amazing survival of individual patients, hypothermic protection after accidental submersion is the exception and can only be expected at water temperatures <5 °C [18]. During the animal experiments the core temperature was reduced to 31 °C while during cardiac surgery <17 °C has been achieved. In neonatal asphyxia the onset of the hypoxic injury begins before hypothermia can be induced. There is no reduction of oxygen consumption in term neonates when the core temperature is above 32 °C [3] and so additional time is needed after the insult to achieve the low body temperatures that may be protective. The studies of hypothermia in asphyxiated neonates were uncontrolled and it is therefore difficult to use these as a basis for therapeutic advice. At the present time hypothermia cannot be recommended as a therapy aimed at protecting the asphyxiated neonate from hypoxic brain damage.

Though asphyxiated neonates are more vulnerable to cold stress than healthy babies [8], there are no controlled trials of delivery room interventions for thermal protection in these infants. Nevertheless thermal protection has to be an integral part of neonatal resuscitation and the guidelines provided by the American Heart Association and the American Academy of Pediatrics [19] mention prevention of heat loss as the first step of neonatal resuscitation. The recommended action is to place the neonate under a radiant warmer, dry the baby thoroughly and remove wet towels. In addition to these recommendations I would make the following comments:

It is important to switch on the radiant heater early to heat up the mattress before arrival of the infant. About 10 % of the neonate's surface area is in direct contact with the mattress and conductive heat loss can be minimized when there is only a small temperature gradient between the infant and the mattress.

Radiant warmers have been extensively studied for premature infants, but there are no special recommendations for the optimal temperature setting of the heater for term neonates during resuscitation. In the intensive care unit the radiant warmer is usually operated with abdominal skin temperature control with the set-point of the servocontrol at 36.5 °C [20]. This is often not feasible during an ongoing resuscitation, when the probe adheres to the skin poorly or is probably covered by sterile towels during umbilical catheter placement etc. Therefore the radiant warmer may be used in the resuscitation area with a fixed heater output as long as the infant is exposed to the heater for only a limited time. During longer exposure it is important to monitor the infant's core temperature every 15 minutes to prevent accidental overheating [21].

Prevention of Heat Loss in the Very Low Birth Weight Infant

Very low birth weight infants are the most vulnerable group in regard to heat loss and the delivery room is not an optimal thermal environment for these babies. This is because :

1. the room temperature is low,
2. the thermal environment is unstable with drafts that increase convective heat loss,
3. the air surrounding the preterm infant has low humidity leading to high insensible water loss, and
4. often there is no source of warmed and humidified oxygen.

There are no controlled trials of interventions in the delivery room aimed at reducing heat loss in the very low birthweight infant and recommendations for their thermal protection are usually extrapolated from studies done in the NICU.

The delivery room temperature should be as high as possible. This can be helped by heat lamps directed, not towards the infant, but at the walls. These will heat not only the air but the walls as well. The smaller the room, the faster the temperature will rise. As much as possible of the infant's surface should be covered with preheated towels, including the head.

Ventilation with warmed and humidified gas during resuscitation is recommended, but there are no controlled studies. Respiratory heat loss in term neonates on day 1 at 32 °C decreased from 9.5 % to 2.9 % of total heat production when humidity was increased from 20 to 80 % [22]. Warm and humidified gas also prevented the decrease in compliance that is observed after only 10 minutes ventilation with non humidified room air [23]. Cold air on the other hand is an important stimulus for the initiation and sustaining of breathing in the newborn. Newborn lambs born into warm water showed a prolonged respiratory depression and did not start to breathe even after the umbilical cord was cut [24]. Therefore controlled trials are needed on the effect of warm and humidified gas for ventilation in the delivery room.

Body temperature should be measured frequently and the best way is by continuous monitoring with a rectal temperature probe. One should be aware that routine clinical thermometers cannot measure temperatures below 34.4 °C. The infants should be transferred into an optimal thermal environment as soon as possible.

Rewarming of Hypothermic Neonates

In the classical descriptions of neonatal hypothermia the authors recommended gradual rewarming [25] without having done any trials. Slow rewarming is achieved by setting the incubator temperature only slightly higher than the infant's temperature and then readjusting the air temperature upwards every few hours. This process takes 12-18 hours [26] and keeping the temperature difference between infant and environment <1 °C was thought to prevent a high oxygen demand before a stable circulation was established.

There are three larger studies on fast rewarming in hypothermic neonates. Tafari [27], in the only controlled study, allocated 30 low birthweight infants (0.81-2.46 kg), admitted to the pediatric university hospital in Addis Ababa at an age of 9-166 h with a rectal temperature <32 °C, either to gradual or to fast rewarming with the incubator temperature set to 36 °C from the beginning. He found the same mortality in both groups (42.9 % in the gradual and 37.5 % in the fast rewarmed group) and concluded from his results that fast rewarming did not have any disadvantages. All infants had intravenous access from the beginning via an umbilical venous catheter and received 100 ml/kgd of 10 % glucose. The interesting additional finding was, that regardless of the mode used, infants who received an extra 20 ml/kg of normal saline at the beginning of rewarming had a better outcome.

Racine [26] reports on 38 neonates in Bethlehem with a weight between 1 and 2.9 kg who were admitted at an age of 1 day to 3 month with a rectal temperature <33 °C. The first 23 infants were treated by gradual and the other 15 by fast rewarming with the incubator temperature set at 37 °C. He observed a decreased mortality with fast rewarming (33 % versus 70 %). The authors note however that there was a treatment difference between the groups that had not been controlled for. The fast rewarmed infants received intravenous fluids 10 to 15 hours earlier than gradual rewarmed infants because iv access was more easily achieved in these infants.

Kaplan [27] reports on 16 neonates with a birth weight between 0.98 and 3.38 kg who were admitted to hospital in the first days of life with a rectal temperature <34 °C. All were treated with fast rewarming in an incubator set at 37 °C. He had no control group but reported a mortality of 19 % which was lower than previously reported in the literature. He gives no information about fluid therapy.

From these studies it can be concluded that fast rewarming does not increase mortality when compared with gradual rewarming, but further studies are needed to demonstrate a beneficial effect. We can generate the hypothesis that early intravenous maintenance fluids and volume replacement therapy decreases mortality in hypothermic neonates. Perhaps the major advantage of fast rewarming is that it facilitates early intravenous access.

In all these studies external rewarming with heated air in an incubator was used to treat hypothermia. There are only case reports about core rewarming in neonates [28]. This is achieved by giving heated gastric and bladder lavages every 30 minutes followed by warmed (36-40 °C) iv fluids as soon as intravenous access is achieved. Core rewarming is increasingly used in pediatric intensive care to rewarm severely hypothermic patients [18] because external rewarming in severe hypothermia can, by inducing peripheral vasodilatation, lead to the dumping of cold acid blood into the central circulation which will cause a secondary temperature drop, rewarming shock and arrhythmias.

Summary

The thermal environment and methods of preventing heat loss in the delivery room need updating because new postdelivery care practices (skin-to-skin care) and a new patient population (very low birth weight infants) have evolved since the classical studies. The best way to rewarm hypothermic neonates is still not known. Skin-to-skin care, placing the naked infant on the mother's chest immediately after delivery, was introduced to promote mother-infant bonding. Healthy term neonates have the same body temperatures with skin-to-skin care as with traditional wrapping in a warm towel, but the former possibly requires greater heat production by the baby. Up to 50 % of very low birth weight infants have body temperatures <36.5 °C in the delivery room and therefore more attention needs to be paid to heat conservation. The ventilation with warm and humidified gas, which is useful in the later care of VLBW infants, is a technique which cannot simply be transferred to the delivery room, because cold air is a necessary stimulus for the onset of respiration. Fast rewarming is not harmful to hypothermic neonates, but its benefit has not been proven in a controlled study. The correct fluid therapy during rewarming has not been established. Together with further research in all these areas, continuous education and training are necessary to ensure the implementation of heat-conserving measures in daily clinical practice.

Implications for Practice

1. Rewarming of Hypothermic Neonates

We suggest the following recommendations for the treatment of hypothermic neonates.

1. Mild hypothermia (32-35 °C)

   - External rewarming at environmental temperature of 37 °C
   - Early iv access

2. Moderate hypothermia with intact circulation (30-32 °C)

   - External rewarming at environmental temperature of 37 °C
   - Early iv access
   - Give warm iv fluids

3. Severe hypothermia (<30 °C)

   - Core rewarming by heated gastric or colonic lavages
   - Try to obtain iv access as early as possible (umbilical vein!)
   - Give warm iv fluids (36 °C-40 °C)
   - Give volume replacement therapy
   - External rewarming at environmental temperature of 37 °C
   - Monitor blood gases, blood glucose, coagulation,

2. Prevention of Heat Loss in the Delivery Room

The following pieces of equipment in addition to the radiant heater are necessary for optimal thermal management:

- additional heat lamps to heat up the resuscitation area when a VLBW infants is expected. Heating up is facilitated if the room is small.
- Thermometer to measure the room temperature
- Warm towels in sufficient quantity. An incubator for microbiologic specimens in the resuscitation area can be conveniently used to heat and store these towels.
- Heater and humidifier for gas used during ventilation of neonates during resuscitation.
- Low reading thermometers to measure body temperature
- Rectal temperature probe for continuous recording of core temperature

3. Prevention of Heat Loss in Healthy Term Neonates, in Asphyxiated Neonates and in Very Low Birth Weight Neonates

- Healthy term neonates
  Dry thoroughly
  In delivery room : Place skin to skin on mother and cover with towel or wrap in warm towel
  For transfer : Wrap in warm towel

- Additional measures for asphyxiated neonates
  Place under radiant warmer with preheated mattress
  When a fixed heater output is used, body temperature must be monitored at least every 15 minutes

- Additional measures for VLBW neonates
  Preheat resuscitation area, minimize drafts
  Cover with preheated towels, include the head
  Provide warm and humidified gas
  Frequently monitor body temperature
  Early transfer to an optimal thermal environment

Recommendations for Future Research

To improve the prevention of heat loss after delivery needs educational and research efforts.
Educational efforts are necessary to make medical staff, nursing staff and midwifes familiar with measures to reduce heat loss and to ensure implementation by training and documentation of heat conserving measures and body temperatures during care of a newborn infant.

Research efforts are necessary in the following areas:

1. Controlled trial of warm and humidified gas for ventilation during resuscitation of VLBW infants.

2. Controlled trial of warm intravenous fluids during resuscitation of VLBW infants

3. Controlled trial of providing intravenous fluids and volume replacement as early as possible during rewarming of hypothermic neonates

4. Can cold stimulation be used for the initiation and the sustaining of breathing

5. Can hypothermia protect against hypoxic-ischaemic injury?

6. Evaluation of educational programmes for training staff to prevent heat loss

References:

1. Brueck K (1961) Temperature regulation in the newborn infant. Biol Neonate 3:65-119

2. Hey EN (1970) The optimum thermal environment for naked babies. Arch Dis Child 45:328-334

3. Adamson K, Gandy DM, James LS (1965) The influence of thermal factors upon oxygen consumption of the newborn human infant. J Pediatr 66:495-508

4. Dahm LS, James LS (1972) Newborn temperature and calculated heat loss in the delivery room. Pediatrics 49:504-513

5. Christensson K, Siles C, Moreno L, Belaustequi A, De La Fuente P. Lagerkrantz H, Puyol P, Winberg J (1992) Temperature, metabolic adaptation and crying in healthy full-term newborns cared for skin-to-skin or in a cot. Acta Paediatr 81:488-493

6. Berliner Neonatalerhebung

7. Johanson RB, Spencer SA, Rolfe P, Jones P, Malla DS (1992) Effect of postdelivery care on neonatal body temperature. Acta Paediatr 81:859-863

8. Gandy GM, Adamson K, Cunningham N, Silverman WA, James LS (1964) Thermal environment and acid-base homeostasis in human infants during the first few hours of life. J Clin Invest 43:751-758

9. Christensson K, Siles C, Belaustequi A, DeLaFuente P, Lagerkrantz H, Puyol P, Winberg J (1993) Lower body temperatures in infants delivered by cesarean section than in vaginally delivered infants. Acta Paediatr 82:128-131

10. Ryser G, Jequier E (1972) Study by direct calorimetry of thermal balance on the first day of life. Eur J Clin Invest 2:176-187

11. Moore RE (1959) Oxygen consumption and body temperature in newborn kittens subjected to hypoxia and reoxigenation. J Physiol 149:500-508

12. Siebke H, Breivik H, Rod T, Lind B (1975) Survival after 40 minutes' submersion without cerebral sequelae. Lancet 1:1275-1277

13. Yager JY, Vannucci RC (1991) Protective effects of mild hypothermia on hypoxic-ischemic brain damage in the immature rat. Pediatr Res 29;366A

14. Kurth CD, Steven JM, Nicolson SC, Chance B, Delivoria-Papdopoulos M (1992) Kinetics of cerebral deoxygenation during deep hypothermic circulatory arrest in neonates. Anesthesiology 77:656-661

15. Westin B., Miller J, Nyberg R, Wedenberg E (1959) Neonatal asphyxia pallida treated with hypothermia alone or with hypothermia and transfusion of oxygenated blood. Surgery 45:868-879

16. Miller J, Miller F, Westin B (1964) Hypothermia in the treatment of asphyxia neonatorum. Biol Neonate 6:148-163

17. Cordey R, Chiolero R, Miller J (1973) Resuscitation of neonates by hypothermia: report on 20 cases with acid-base determination on 10 cases and the long-term development of 33 cases. Resuscitation 2:169-181

18. Corneli HM (1992) Accidental hypothermia. J Pediatr 120:671-679

19. American Heart Association and American Academy of Pediatrics (1990) Textbook of neonatal resuscitation.

20. Malin SW, Baumgart S (1987) Optimal thermal management for low-birth weight infants nursed under high-powered radiant warmers. Pediatrics 79:47-54

21. Mayfield SR, Bhatia J, Nakamura KT (1984) Temperature measurement in term and preterm neonates. J Pediatr 104;271

22. Sulyok E, Jequier E, Prod'hom LS (1973) Respiratory contribution to the thermal balance of the newborn infant under various ambient conditions. Pediatrics 51:641-650

23. Greenspan JS, Wolfson MR, Shaffer TH (1991) Airway responsiveness to low inspired gas temperature in preterm neonates. J Pediatr 118:443-445

24. Harned HS, Herrington RT, Ferreiro JI (1970) The effects of immersion and temperature on respiration in newborn lambs. Pediatrics 45:598-602

25. Bower BD, Jones LF, Weeks MM (1960) Cold injury in the newborn: a study of 70 cases. Brit Med J 1:303-309

26. Racine J, Jarjoui E (1982) Severe hypothermia in infants. Helv Paediatr Acta 37:317-322

27. Tafari N, Gentz J (1974) Aspects of rewarming newborn infants with severe accidental hypothermia. Acta Paediatr Scand 63:595-600

28. Kaplan M, Eidelman AI (1984) Improved prognosis in severely hypothermic newborn infants treated by rapid rewarming J Pediatr 105:470-474

29. Thompson DA, Anderson N (1994) Successful resuscitation of a severely hypothermic neonate. Ann Emerg Med 23:1390-1393

Chapter 17

Transportation of Sick Newborns in a Cold Environment

GUNNAR SEDIN / KARL HENRIK FRIDBLOM

Introduction

Seriously ill and very preterm infants who have to be transported to a regional or national centre for treatment and/or intensive care, need well planned transport with appropriate equipment, well trained personnel and a vehicle that fulfils all requirements for the transport of a newborn infant in an incubator. Before transportation, the condition of the infant should be stabilized so that the gas exchange and oxygenation are adequate and the body temperature normal (36.0 - 37.0 °C) [1, 2]. Any electrodes for monitoring should be applied, catheters for sampling and infusion should be introduced, and fluid supply started to maintain an appropriate energy balance. All equipment for the transport should be checked carefully and drug administration planned before the start of the transport. The transport should be arranged so that it is both safe and comfortable for the infant, with vibration and noise kept as low as possible.

One of the risk factors in neonatal transports, especially in a cold climate, is that the heat supply might not be sufficient or the heat loss too high to make it possible to maintain an appropriate incubator air temperature and to keep the body temperature of the infant normal. As cooling of the infant during transport may be life threatening [3, 4, 5], maintenance of the infant's thermal balance should be given high priority in the planning of the transport and in the choice of equipment and vehicle [6, 7, 8].

Routes of Heat Exchange and Heat Supply

The heat exchange between the infant's skin and the environment takes place through conduction (H_{cond}), evaporation (H_{evap}), radiation (H_{rad}) and convection (H_{conv}), and that between the infant's respiratory tract and the environment through evaporation (H_{evap-r}) and convection (H_{conv-r}) [9, 10, 11, 12, 13, 14]. There are a number of circumstances that increase the risk of a fall in body temperature during transport either due to excessive heat losses or to insufficient heat supply, e.g. an insufficient power supply to, or heating capacity of, the incubator, unfavourable environmental conditions outside or inside the incubator, and disturbances during the transport, such as medical complications that need intervention.

Preparation for Transport

Disturbances in heat balance during transport can be avoided by seeing that the transfer is planned and the equipment well adjusted and functional. Before placing the infant in the transport incubator, a check should be made to see that it is heated and, if provided, that the humidifier is filled with sterile water. The battery of the transport incubator must be well charged before transport. The transport team should have information concerning available sources of electrical supply and, if possible, the battery of the transport incubator should be used only to heat the incubator when no alternative supply is available in the sending and receiving hospital or in the vehicles. The volume of oxygen and compressed air that will be needed has to be estimated and taken with the transport. Equipment for monitoring ECG, respiration, oxygen saturation and temperature should always be present and when necessary, additional equipment for monitoring of arterial blood pressure, $tcPO_2$ and $tcPCO_2$.

Insufficient Heat Supply

This may occur when the battery is not well charged before the start of the transport, or when it has been used for too long. The batteries used in transport incubators should be non-spillable and placed in a separate box. They must be approved by the Civil Aviation Authority and should give 60 Ah (12 V) [6].

In a very cold climate the generators of ground ambulances occasionally have insufficient capacity to provide electrical power for the ambulance itself, including its lights, heating, signal systems and so on, in which case the delivery of electrical power to the incubator will also be insufficient. This can be avoided if the ambulances are provided with high capacity generators [6]. In helicopters and winged ambulance planes, supply of electrical power to the incubator is not usually a problem. Recently one of the airlines (SAS) has installed 220 V (AC) in all its aircraft to meet the requirements for long distance transport of incubators and other medical equipment. This follows a request raised by The Swedish Study Group for Neonatal Incubator Transports [6, 8] and this measure is also approved by the Study Group on Perinatal Transports of the European Association of Perinatal Medicine [8].

Environmental Conditions Outside the Incubator

The surface of the incubator exchanges heat with its environment. When the environmental air is very cold, the temperature of the transparent parts of the incubator walls will decrease dramatically. Also, the air flow velocity outside the incubator will influence the rate at which the temperature of the incubator walls changes. This will then have secondary effects on the air temperature (T_{air}) within the incubator. High air flow velocities created by the helicopter rotors can also increase the heat loss from transparent walls. Examples are given below.

Environmental Conditions Within the Incubator:

In the transport incubators the heat source is positioned so that it heats the compartment under the bed, parts of the walls or the ceiling. If only one wall is heated, that wall will have a much higher temperature than the other surfaces facing the infant. This difference will be exaggerated if some walls are less well insulated than others. The incubator air is heated from the same heat source and differences between the incubator walls will also cause an uneven distribution of air temperature (T_{air}).

The incubator air is usually very dry, as very few transport incubators are equipped with a humidification system. The respirators used in transport incubators are usually not equipped with a heating and humidification system and the inspired gas is therefore both cold and dry.

Heat Exchange between the Infant and the Environment

Heat Exchange through the Skin

The different modes of heat exchange between the infant's skin and the environment in the transport incubator will naturally be influenced by the temperature of the bed (affecting H_{cond}), the humidity of the incubator air (H_{evap}), the temperature of the incubator walls (H_{rad}) and the temperature of the incubator air (H_{conv}) [9, 10, 12]. The insulation of the mattress in transport incubators is usually good and the losses of heat through conduction (H_{cond}) will be low.

Heat Exchange through Evaporation

As most transport incubators are not equipped with a humidification system, dry air in the incubator will result in high transepidermal water loss (TEWL) and a high evaporative loss of water and heat (H_{evap}) from the skin surface [8, 12, 13]. Humidification of the air in transport incubators is given high priority in the requirements listed by the Swedish Study Group on Neonatal Incubator Transports [6]. Humidification of air may make it necessary to keep the inner walls of the incubator warm to avoid condensation.

Heat Exchange through Radiation

H_{rad} depends on the temperatures of the surfaces facing the infant (T_{wall}) and the skin temperature (T_{skin}) [9, 10, 12]. The surfaces that provide heat to the transport incubator are usually so warm that they cause a gain of heat through radiation, while those surfaces that are transparent have a much lower temperature and cause a loss of heat from the infant. Installation of transparent double walls in transport incubators would certainly lower the heat loss through radiation through these parts of the incubator (6, see example below).

Heat Loss through Convection

H_{conv} depends on the air and skin temperatures [9, 10, 12]. The air temperature (T_{air}) is unevenly distributed within a transport incubator, partly because there is no fan circulating the air around the infant. The air temperature in the incubator will also fall dramatically if the small ports or the main opening of the incubator has to be kept open for nursing or medical interventions during the transport.

Heat Exchange through Conduction

H_{cond} is usually low if the mattress has a good insulation. Heat can be gained through this route of heat exchange but heated mattresses have been considered to introduce too high a risk of burns, especially in infants with poor peripheral circulation. If a mattress can be safely warmed to a temperature which will not cause burn injury, heat loss through this route can be eliminated and turned into heat gain.

Means to Achieve Lower Heat Exchange

Large differences between the temperatures of the incubator surfaces and an uneven air temperature distribution within the incubator might cause some discomfort to the infant, who might feel a draught. To be able to maintain a normal body temperature it is essential to cover any open lesions, e.g. a myelomeningocoele, omphalocoele etc, with a protective dressing covered with a plastic wrap to reduce radiant and evaporative heat losses. It is also possible to decrease the radiant heat loss during transport by use of a silver swaddler or by using a cover with plastic cellular wraps or a blanket. These can easily be removed, which is more difficult with clothing that may limit the access to the critically ill infant and reduce the possibility of observing skin colour.

Heat Exchange through the Respiratory Tract

This heat exchange takes place through convection (H_{conv-r}) and evaporation (H_{evap-r}). The exchange of heat through convection depends on the difference between the temperature of the inspired (T_I) and expired (T_E) air and the ventilation volume (V). For further information see refs. [9, 14 and 15]. In infants nursed in a transport incubator H_{conv-r} will be low when the incubator air temperature is maintained at an appropriate level. Respirator treatment using a cold gas for inspiration will result in considerable heat loss.

The evaporative heat exchange through the respiratory tract depends on the difference in water content between the expired and inspired air. H_{evap-r} will therefore be higher if the inspired gas is not humidified. As the water content of the air also depends on the temperature, the gas to be inspired by the infant during a transport should be both warmed and humidified to attain a high vapour pressure and a low H_{evap-r}. Thus, the heat loss from the respiratory tract (H_{conv-r} and H_{evap-r}) can be lowered by warming and humidifying the gas which the infant inspires.

Heat Exchange between the Transport Incubator and the Environment.

Below are examples of situations in which changes in the temperature or air flow velocity will alter the heat exchange within the incubator environment.

Cold Environment:

The first example shows how changes in the temperature of the room will influence those of the different walls of the transport incubator (Dräger 5400). Measurements were made during a 55-minute period in room air and during a longer period in a room with a temperature of 7 °C. During the time at the lower temperature two small ports were opened in the front wall of the incubator. As seen in figure 17.-1, the temperature of the back wall of the incubator, where the heat source is placed, was much higher than that of the front wall. When the incubator was transferred to a room with a temperature of 7 °C, the fall in temperature was also much more marked at the inside of the front wall, which was transparent. Opening of the two small ports caused a further decrease in the wall temperatures and the incubator air temperature also decreased markedly. After the ports were closed it took 40 minutes for the air temperature, and 25-40 minutes for the wall temperatures, to return to a level equal to that before the ports were opened.

Table 17.-1 shows the temperatures, humidity and calculated heat exchange 30, 120, 190 and 240 minutes after the start of the measurements presented in figure 17.-1. The temperature of the dummy used in the incubator decreased, but in these calculations the skin temperature (T_{skin}) was kept constant at 36.5 °C. Temperatures are given both for the front and the back wall, and the radiative heat exchange (H_{rad}) was calculated between these walls and an infant born at 25 weeks of gestation, using data from a previously published paper [11] on TEWL and skin temperatures.

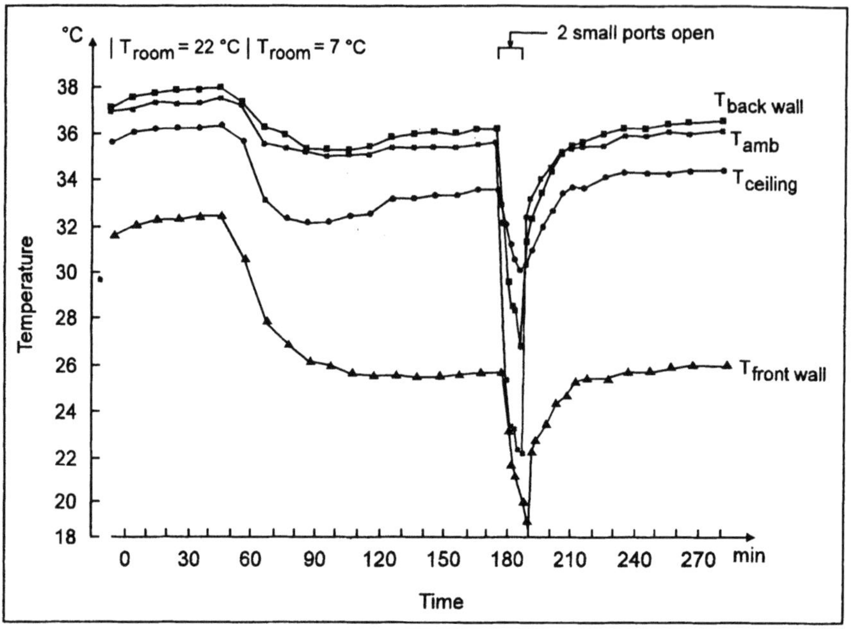

Fig. 17.-1 : Temperatures of transport incubator air and incubator walls during measurements at two different room temperatures and during opening of the small ports of a Dräger 5400 transport incubator.

As seen in table 17.-1, the relative humidity (RH_{amb}) and the water vapour pressure (P_{H2Oamb}) were low and the evaporative heat loss, for an infant of this gestational age, was very high. There were marked differences in heat exchange with the environment between the surfaces of the infant facing the front compared with those facing the back wall. Thirty minutes after the start of measurements there was a radiative heat loss from the infant from surfaces facing the front wall, but a heat gain for those facing the back wall. These differences in radiative heat exchange between the two sides of the infant increased when the temperature of the front wall decreased (see e.g. 120 and 190 minutes). Heat loss through convection was initially low, but increased markedly when the ports of the front wall were open. The heat loss through evaporation remained high throughout the measurement period.

Tab. 17.-1 : Temperatures and humidity and calculated heat exchange in a transport incubator in a room air temperature of 22.0 °C and 7.0 °C (55 min after start of measurement). T_{skin} is considered to be constant at 36.5 °C. H_{evap} is calculated from data published in Acta Paediatrica Scand 68, 795, 1979 and 71, 191, 1982 for an infant born after 25 weeks of gestation.

Time min	T_{room} °C	T_{amb} °C	T_{wall} °C Front/Back wall	T_{skin} °C	RH °C	P_{H2O} kp	H_{evap} w/m²	H_{rad} w/m² Front/Back wall	H_{conv} w/m²
30	22.0	37.3	32.3 / 37.8	36.5	11	5.4	55.3	10.4 / -4.0	-2.2
120	7.0	35.0	25.4 / 35.3	36.5	8	4.2	57.3	18.0 / 7.0	4.0
190	7.0	32.1	21.9 / 31.0	36.5	9	4.1	56.7	24.6 / 16.8	39.4
240	7.0	35.5	25.3 / 35.8	36.5	7	3.9	58.0	18.1 / 6.4	2.7

Evaporative heat loss from the respiratory tract of the infant will be high due to the dry air. Both evaporation and convective heat losses will increase as the air temperature drops.

These measurements show clearly that in an environment with a temperature of 7 °C, heat loss from the surface of the incubator cannot be compensated for by an increased heat supply from the incubator. The initial decrease in temperatures 50 minutes after the start of the measurements would probably have already caused a significant drop in body temperature in a preterm infant. Opening of the ports would cause a further decrease. Improved warming of the incubator air and elevation of the temperature of the walls, especially the transparent front wall, would have decreased the heat loss markedly. A safe and well-controlled warming of the mattress to a surface temperature never exceeding 38 °C [17, 18] could make it possible to eliminate loss of heat through conduction and also contribute to the heating of the incubator air. The temperature of the inner side of the transparent wall can be increased by using double shells, by putting a cover over the incubator or by heating the wall. Alternatively, radiant heat can be provided from parts of the walls and the ceiling [16].

High Air Velocity

The temperature of the ceiling of the transport incubator changed when, after 40 minutes of measurement, the incubator was placed just outside the helicopter ready to be loaded (Figure 17.-2). The incubator initially had a thick cover placed over it providing good insulation. When the helicopter rotor was started there was a small decrease in temperature of the ceiling and when the cover was removed the temperature dropped immediately from 38 to 33 °C. This shows that close to a helicopter the incubator temperature can be maintained reasonably well if there is an insulating cover but the high air velocity, at a temperature of 14 °C, will markedly influence the temperature of the incubator ceiling.

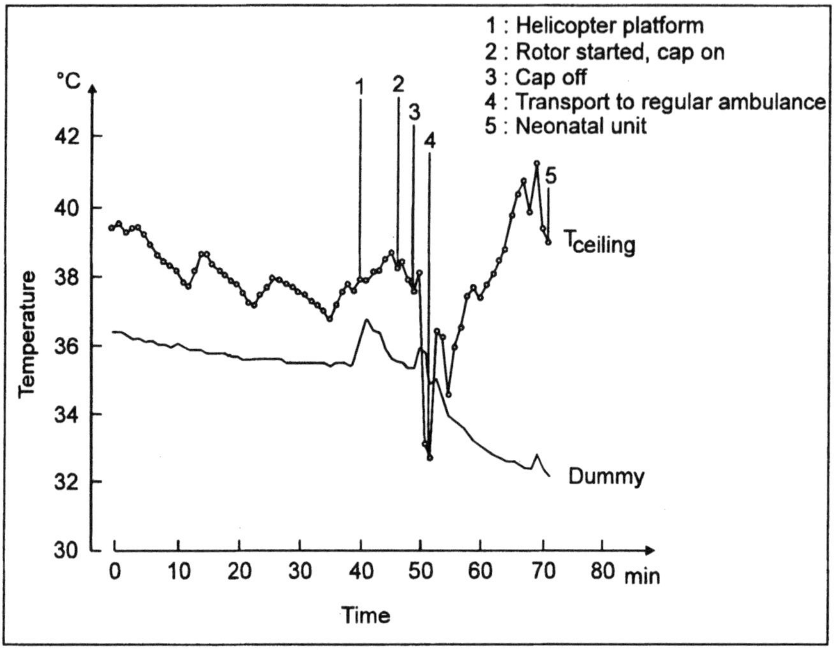

Fig. 17.-2 : Temperature of the ceiling of a transport incubator placed close to a helicopter. The incubator was covered with a cap, which was removed after the rotor had started to move.

To conclude, humidification and warming of the air in transport incubators would markedly lower the heat loss from evaporation. An evenly distributed inner wall temperature would decrease the heat loss through radiation and make the incubator more comfortable for the infant. To provide an environment that will allow the infant's thermal balance to be maintained during transport, the incubator air temperature and that of the walls must be maintained even when the environment changes in regard to temperature and air flow velocities. Humidification and warming of the gas that the infant inspires will lower the heat loss from the respiratory tract from both evaporation and convection.

All improvements and additions to functions of transport incubators should be made without increasing the total weight. New or altered functions must be tested to check that they do not have any unexpected side effects which might disturb the function of the incubator, other medical equipment, the vehicle, the communication system, or personnel needed for medical care or for the transport vehicle.

Abbreviations

H_{cond}, heat exchange through conduction between the skin and the environment (W/m^2)
H_{evap}, heat exchange through evaporation from the skin surface (W/m^2)
H_{rad}, heat exchange through radiation between the skin and environment (W/m^2)
H_{conv}, heat exchange through convection between the skin and the environment (W/m^2)
H_{evap-r}, heat exchange through evaporation from the respiratory tract
H_{conv-r}, heat exchange through convection in the respiratory tract
P_{H2Oamb}, water vapour pressure of ambient air
RH_{amb}, relative humidity of ambient air
TEWL, transepidermal water loss (g/m^2 x h)
T_{bed}, temperature of bedding (mattress) (K)
T_{skin}, skin temperature (K)
T_{amb}, temperature of ambient air (K)
T_{wall}, temperature of the wall of an incubator or a room (K)
T_E, temperature of expired gas (C)
T_I, temperature of inspired gas (C)
V, ventilation volume (l/unit time)

Acknowledgement: This paper is based on studies supported by the Swedish Medical Research Council (project 19X-04998) and on information collected by the Swedish Study Group on Neonatal Incubator Transports. The authors are very grateful to Mrs Barbro Kjällström and Mrs Gertie Ericson for their excellent technical assistance during the measurements.

References:

1. Blake AM, McIntosh N, Reynolds EOR, Andrew DST (1975) Transport of newborn infants for intensive care. British Medical Journal 4:13-17

2. Malhotra AK, Deorari AK, Paul VK, Bagga A, Singh M (1992). A new transport incubator for primary care of low birth weight babies. Indian Pediatrics 29:587-593

3. Silverman WA, Fertig JW, Berger AP (1958) The influence of the thermal environment upon survival of newly born premature infants. Paediatr 22:876-885

4. Chance GW, O'Brien MJ, Swyer PR (1973) Transport of sick neonates, 1972: An unsatisfactory aspect of medical care. Can Med Assoc J 109:847-851

5. Chance GW (1978) Thermal environment in transport. Temperature regulation and energy metabolism in the newborn. Ch 9, 227-240

6. Requirements raised by the Swedish Study Group on Neonatal Incubator Transports. Swedish Health Care Standard Institution, Stockholm, Protocol HSS-SIS 930907 and CEN/TC 239/N87 and CEN/TC 239/ WG3/N152

7. Guidelines for air and ground transport of neonatal and pediatric patients. American Academy of Pediatrics 1993

8. Sedin G, Gillblom P, Leding L (1994) Interfacing of transport incubators in different vehicles, 14th European Congress of Perinatal Medicine, June 5-8, Helsinki, Abstract 48

9. Houdas Y, Ring EFJ (1982) Human body temperature. Its measurement and regulation. Plenum press, New York and London

10. Hammarlund K, Sedin G (1982) Transepidermal water loss in newborn infants. VI. Heat exchange with the environment in relation to gestational age. Acta Paediatr Scand 71:191-196

11. Hammarlund K, Sedin G (1979) Transepidermal water loss in newborn infants. III. Relation to gestational age. Acta Paediatr Scand 68:795-801

12. Cooney DO (1976) Biomedical engineering principles. An introduction to fluid, heat, and mass transport processes. New York and Basel: Marcel Dekker Inc., 93-155

13. Hey EN, Mount LE (1967) Heat losses from babies in incubators. Arch Dis Child 42:75-84

14. Sedin G. Neonatal heat transfer, routes of heat loss and heat gain. **This volume**, chapter XX

15. Riesenfeld T, Hammarlund K, Sedin G (1990) The effect of a warm environment on respiratory water loss in fullterm newborn infants on their first day after birth. Acta Paediatr Scand 79:893-898

16. Lavaud J, Chabernaud JL, Barbier ML, Fevrier YM, Johanet S, Hobeika G, Lode N, Andre P (1992) (second edition, first edition 1985). Réanimation et transport pédiatriques. Ed: Masson, 120 Boulevard, 75280 Paris Cedex 06. ISBN 2-225-82529-7

17. Tunell R (1994) Treatment of sick new-borns with heated mattresses. A guideline for: temperature control, temperature and optimal thermal environment for sick and low birth weight neonates. **This volume**, chapter 14

18. Koch J (1994) Physical properties of the thermal environment. A guideline for: temperature control, temperature and optimal thermal environment for sick and low birth weight neonates. **This volume**, chapter 10

MIX
Papier aus verantwortungsvollen Quellen
Paper from responsible sources
FSC® C105338

If you have any concerns about our products,
you can contact us on
ProductSafety@springernature.com

In case Publisher is established outside the EU,
the EU authorized representative is:
Springer Nature Customer Service Center GmbH
Europaplatz 3, 69115 Heidelberg, Germany

Printed by Libri Plureos GmbH
in Hamburg, Germany